EEA Report | No 5/2009

Ensuring quality of life in Europe's cities and towns

Tackling the environmental challenges driven by European and global change

European Environment Agency

Cover design: EEA
Cover photo: © Birgit Georgi
Left photo: © Jens Rørbech
Right photo: © Jan Gehl and Lasse Gemzøe
Layout: EEA/Pia Schmidt

Information about the European Union is available on the Internet. It can be accessed through the Europa server (www.europa.eu).

Luxembourg: Office for Official Publications of the European Communities, 2009

ISBN 978-92-9167-994-2
ISSN 1725-9177
DOI 10.2800/11052

Environmental production
This publication is printed according to high environmental standards.

Printed by Schultz Grafisk
— Environmental Management Certificate: ISO 14001
— IQNet - The International Certification Network DS/EN ISO 14001:2004

— Quality Certificate: ISO 9001: 2000
— EMAS Registration. Licence no. DK - 000235
— Ecolabelling with the Nordic Swan, licence no. 541 176

Paper
RePrint — 90 gsm.
CyclusOffset — 250 gsm.
Both paper qualities are recycled paper and have obtained the ecolabel Nordic Swan.

Printed in Denmark

European Environment Agency
Kongens Nytorv 6
1050 Copenhagen K
Denmark
Tel.: +45 33 36 71 00
Fax: +45 33 36 71 99
Web: eea.europa.eu
Enquiries: eea.europa.eu/enquiries

Contents

Acknowledgements

This report was written and compiled by:

- the European Environment Agency (EEA): Birgit Georgi, Dorota Jarosinska, Almut Reichel, Jaroslav Fiala, Anke Lükewille, Colin Nugent, Josef Herkendal, Stéphane Isoard, Gorm Dige, Elena Cebrian Calvo, David Delcampe, Peder Gabrielsen;
- EEA Topic Centre on Land Use and Spatial Information (ETC-LUSI): Jaume Fons, David Ludlow, Stefan Kleeschulte;
- ICLEI-Local Governments for Sustainability: Holger Robrecht, Cristina Garzillo ;
- the Network of European Metropolitan Regions and Areas (METREX): Vincent Goodstadt, Will French;
- Energie-Cités: Kristina Dely;
- Council of European Municipalities and Regions (ÇEMR): Marie Bullet, Boris Tonhauser;
- Union of Baltic Cities (UBC) Environment and Sustainable Development Secretariat: Anna Granberg, Niina Salonen;
- Ambiente Italia S.r.l. — Research Institute: Maria Berrini, Lorenzo Bono;
- Architects' Council of Europe (ACE): Adrain Joyce;
- Joint Research Centre of the European Commission — Institute for Environment and Sustainability: Carlo Lavalle;
- Netherlands Environmental Assessment Agency: Judith Borsboom, Rob Folkert, Stefan Berghuis, Ton Dassen.

Additional contributors: EUROCITIES Environment Forum: Eva Baños, Jan Meijdam, Henk Wolfert, information on noise (Section 2.4) and climate change (Section 2.5); Beate Arends (Province of South-Holland) and Simone Goedings (Association of Dutch Municipalities [VNG] for CEMR), information on air pollution (Section 2.4).

The report team also wishes to thank the many further experts consulted throughout the development of this report, in particular: Michelle Dobré (University of Caen-Normandy, researcher in Centre Maurice Halbwachs); Pierre Laconte (Foundation for the Urban Environment, Member of the EEA Scientific Committee); Sivia Brini, F. Moricci, A. Chiesura, and M.C. Cirillo (all ISPRA Italia); Giovanni Fini (Municipality Bologna); Antonín Tym (Healthy Cities Czech Republic); Daniel Skog (Municipality Malmö); Florian Ismaier (Municipality Karlsruhe); Eduardo Miera (URBAN Programme San Sebastián-Pasaia); Toni Pujol (Municipality Barcelona); Karen Hiort (Municipality Berlin); Monika Gollnick (Municipality Ludwigshafen); Dieter Teynor (Municipality Mannheim); Torun Israelsson (Municipality Växjö); Thierry Lavoux (French Ministry of Sustainable Development & Environment); Josiane Lowy (Conseillère régionale de la Région Basse Normandie); Teodora Brandmueller, Corinne Hermant-de Callataÿ and Marcel Rommerts (European Commission); Simone Reinhart (European Parliament); Didier Vancutsem (International Society of City and Regional Planners [ISOCARP]); Hedwig Verron and Christoph Erdmenger (Umweltbundesamt, Dessau); Tatiana Bosteels (Hermes Real Estate Investment Management Limited, London).

Finally, we would like to thank the Swedish Environment Ministry for its financial support.

The report was coordinated and edited by Birgit Georgi and Ronan Uhel (EEA), supported by David Ludlow (University of the West of England, Bristol) and Michelle Dobré (University of Caen-Normandy, researcher in Centre Maurice Halbwachs).

Preface

In May 2008, the Council of Europe's Congress of Local and Regional Authorities captured the concerns and desires of urban policy-makers and citizens in the title of its new European Urban Charter: Manifesto for a new urbanity. Like numerous other international and European charters, conventions and declarations, the manifesto describes with some apprehension the 'unprecedented environmental, democratic, cultural, social and economic challenges' facing urban centres and their inhabitants.

Our report on quality of life in Europe's cities and towns reiterates these concerns but also unravels the many apparent paradoxes of urban development and the sometimes perplexing realities of urban Europe today. The report defines a vision for progress towards a more sustainable, well-designed urban future, and in doing so inevitably raises many questions:

- why call for a new urbanity at a time when Europeans' living standards, notwithstanding the current global economic downturn, have on average and over decades progressively risen?
- why call for a new urbanity when it is evident that urban governance measures have improved living conditions?
- why call for a new urbanity to be delivered by our political leaders, the construction sector and ordinary citizens, when the vast majority of urban areas have benefited from this new prosperity?

The simple answer to these apparent paradoxes is evident in the many concerns expressed by the vast majority of policy-makers, professionals and civil society. They point out that the current urban model delivers higher living standards and prosperity but fails to deliver 'quality of life'. Unsurprisingly, the complex interaction between the many determinants of quality of life means that efforts to promote one element can have unexpected impacts elsewhere. However, understanding these apparent paradoxes is vital to realising the vision of a vibrant urban future in which economic, social and environmental aspirations can be delivered concurrently.

The notion of 'quality of life' normally implies broad and long-term societal objectives and indicators, which can be at odds with the short-term, sectoral targets that guide much policy-making. With that in mind, the prime aim of this report is to explore the many perceptions of quality of life in order to help define urban problems more clearly, identify options for remedial action and construct evaluations of effectiveness. All these areas are relevant to improving the governance of today's urban realities throughout Europe.

This report highlights the connections between the different dimensions of quality of life and analyzes the inherent causal relationships. These range from clear linkages such as the health benefits of green open space for urban populations to less evident relationships such as the way that individual choice of housing has environmental impacts that affect quality of life. In this way, the report addresses the sustainable design and development of Europe's cities, perceiving environmental quality as a fundamental building block of social well-being and urban quality of life.

Realising the vision of a more sustainable urban future requires mobilising action and resources to reconstruct towns and cities. The aim should be to create new social, cultural and economic foundations that conserve the environmental underpinnings and so offer long-term benefits for Europe's future generations.

With humility, our report is the result of the endeavours and expertise of many individuals collectively representing a number of pan-European organisations and it attempts to cover the many issues inherent to urban complexity. Cities and towns are essentially bodies of coexistence; calls for a new urbanity may thus reflect a shared awareness that fragmented and short-term policies are hindering urban areas from fulfilling that core function.

The authors

What is this report about?

Quality of life is a term broadly used both by the general public and amongst policy-makers. Everyone agrees on its importance, but a definitive meaning cannot be assigned to it — the term can mean many things to many people. In recognition of this diversity of perspectives, a range of partners with distinct backgrounds dealing with urban issues across Europe discussed their views and provided their results in this joint report.

Aims of this report

This report aims to raise awareness of the various perspectives on, and perceptions of, quality of life. It stresses the challenges ahead to ensure quality of life in the long run for all social groups, and the crucial importance of sustainability and the environment as our life-supporting system. The report sheds light on certain aspects of the current quality of life discussions but without attempting to provide a finite scientific definition, as the authors recognise that the many subjective aspects of quality of life do not permit the derivation of an objective, universal definition. It is the role of individuals and political representatives to formulate and agree on a concept for quality of life for their needs and for their purposes. The report aims to illustrate how different conceptions of quality of life influence the quality of life of others, and provides ideas for ways to meet the challenges that lie ahead; and by doing so aims to support individuals and politicians to discover a balanced concept for quality of life compatible with sustainable development.

Urban perspectives

The spatial focus of this report is on cities and towns in Europe. Urban dwellers represent the overwhelming majority of the European population. Cities and towns are therefore the places where, for most people, quality of life is experienced and delivered. However, cities and towns, whilst providing many services for the rural population, also consume rural services. This means that urban and rural areas are strongly interlinked. As a result, quality of life in urban areas also impacts on that in rural areas.

Photo: © Pavel Šťastný

Whilst interconnected with rural areas, cities and towns also interact with each other, and function in regional, national and European frameworks. For example, European policy sets the framework in which national, regional and local governments act. Similarly the impact of local policies, such as the reduction of local greenhouse gas emissions, influences the European situation by reducing overall emissions and so contributing to climate change mitigation. In conclusion, concepts to ensure quality of life in cities and towns need to consider these interlinkages and require the participation of all administrative levels.

Policy focus

Accordingly, this report in particular addresses the concerns of policy- and decision-makers in cities and towns, as well as those at European level who deal directly or indirectly with urban issues. The report also provides useful information and arguments for regional and national authorities and other interested stakeholders and groups, including business, non-governmental organisations (NGOs) and the general public.

In summary, this report aims to raise awareness of the remarkable potential of cities and towns to deliver quality of life, not only for their own populations, but also for all European citizens. In addressing the problems cities and towns face in realising this potential, the report focuses on the network of local, national, European and global interactions and the impacts of global change and other environmental challenges as they impact on quality of life. Finally, the report provides ideas and good practice examples of integrated action, policy responses and governance to tackle the problems and master the challenges.

Ways to read the report

The report offers different opportunities for review according to the specific background, responsibilities and interests of the reader:

- for some, the Chapter 1 overview may be sufficient;
- others may have an interest in the more detailed specification of drivers of change and the challenges faced at the urban level, together with ideas for remedial action set out in Chapter 2;
- Chapter 3 provides specific ideas on the establishment of an integrated policy approach linking thematic areas and all administrative levels as a major initiative to deliver quality of life in a balanced way.

1 Quality of life in European cities and towns

1.1 Quality of life — the urban crossroads of all policies

The desire for quality of life is universal and generates consensus across political and popular arenas. This common goal can assist all responsible agencies and citizens to overcome their differences and coordinate their responses.

Now, more than ever before, Europe's wealth, innovation potential, creativity and talent are centred in its wide range of towns and cities. Quality of life and quality of the environment underpin how well these towns and cities function. Cities are business hubs, attracting investment to create jobs, and provide the focus of service provision and exchange. Urban areas are also the focus of many environmental challenges, where quality of life is determined by a wide mix of socio-economic and political factors. Therefore, our towns and cities are where the interwoven challenges of quality of life and sustainable development must primarily be addressed.

Progress towards quality of life

Undoubtedly, quality of life has improved in many areas over the past 50 years. Today we benefit from more welfare and more living space per person, own more cars, travel more and further in our work and holidays, enjoy luxury goods and live longer. However, in other areas, particularly health, quality of life has deteriorated. For example, there have been marked increases in allergic reactions and lifestyle-related diseases, such as cardiovascular disorders caused by obesity, physical inactivity or stress.

Individual searches for a better quality of life, such as a better quality of domestic living environment, drive urban migrations and urban sprawl. This has unintended negative consequences for society as a whole. Growing consumption is putting our environment under increasing pressure with consequences for quality of life. Excessive energy consumption exacerbates harmful climate change, for example heat waves such as the one in Europe in 2003 caused tens of thousands of premature deaths. Continuing growth in mobility generates more noise and air pollution and increasing land consumption has negative impacts on biodiversity and ecosystems.

There is notable conflict between individual short-term quality of life benefits and collective, longer-term needs for sustainable development that forms the basis for quality of life in the future.

Quality of life is a concern for every social group, but significant inequalities persist; for example, in degrees of exposure to pollution and industrial risks, and access to better living conditions. However, the privileged in society are often able to improve their quality of life, for instance by moving to better neighbourhoods or to the countryside in order to escape from unhealthy conditions.

Political consensus but competing conceptions

Quality of life is a feature of many political (Box 1.1) and scientific agendas. However, because perception of quality of life, particularly in urban areas, differs so much, local policies are often very diverse. The fact that quality of life is rarely adequately defined in official documents only serves to exacerbate the situation, and results in policies that focus on specific areas such as income, housing or local environment, without taking a broader view. This can generate contradictory development paths. For example, prioritising jobs and economic growth to secure quality of life can result in negative environmental impacts.

Similarly, differing perceptions can affect policy-making at government level and result in distinct and different views on the priorities for socio-economic development and diverging recommendations on what, if anything, governments should do in order to promote the quality of life in Europe's cities and regions. The challenge is understand these differences and to formulate a simple definition of quality of life. By doing so, policy-makers will gain public support and be better able to work with all stakeholders to

Box 1.1 Political committments to quality of life

The Treaty on European Union (consolidated version 2008)
The Union's aim is to promote peace, its values and the well-being of its peoples.

Renewed EU Sustainable Development Strategy (SDS) 2006
The overall aim of the renewed EU SDS is to identify and develop actions to enable the EU to achieve continuous improvement of quality of life both for current and for future generations, through the creation of sustainable communities able to manage and use resources efficiently and to tap the ecological and social innovation potential of the economy, ensuring prosperity, environmental protection and social cohesion.

Leipzig Charter on Sustainable European Cities and Bristol Accord
The Charter gives no definition but aims at '... a high quality in the fields of urban design, architecture and environment'. It builds on the Bristol Accord which define sustainable communities as 'places where people want to live and work, now and in the future. They meet the diverse needs of existing and future residents, are sensitive to their environment, and contribute to a high quality of life. They are safe and inclusive, well planned, built and run, and offer equality of opportunity and good services for all'.

EU Thematic Strategy on the Urban Environment
Four out of five European citizens live in urban areas and their quality of life is directly influenced by the state of the urban environment. A high quality urban environment also contributes to the priority of the renewed Lisbon Strategy to 'make Europe a more attractive place to work and invest'.

The Aalborg Charter of European Cities and Towns towards Sustainability
Aims to 'integrate environmental with social and economic development to improve health and quality of life for our citizens'.

agree on a coherent and comprehensive vision of quality of life to support targeted policies.

Tackling the mismatch

The current mismatch between popular conceptions of quality of life now and the longer term needs for sustainability as the basic fundament to quality of life in the future (Box 1.2) is a critical issue. Policies need to distinguish between quality of life that produces demands for general basic needs, for example access to services, and demands arising from individual lifestyles that encourage higher consumption. Policies must be based on an equitable vision of quality of life and balance priorities for today without comprising the global environment and the lives of future generations. Clearly, some aspects of our current ways of life require shifts toward more socially and environmentally oriented priorities and, as a consequence, adaptation to more sustainable lifestyles at both individual and societal level.

A major problem is that this mismatch is rarely transparent. This can undermine the political support necessary to secure both sustainable development and a sustained quality of life. It is therefore vital to raise public awareness of the impacts of the pursuit of short-term quality of life at the expense of longer-term sustainable development.

Unifying quality of life and sustainability

All the above highlights the critical links between environmental sustainability, quality of life and the future success of cities expressed in terms of social and economic as well as environmental factors. The Stern Report (Stern, 2006) on the economics of climate change, for example, demonstrates that the real economic costs of unsustainable living and further climate change are much higher than the cost of investments in climate change mitigation and adaptation. The shift to more sustainable lifestyles is therefore not simply a matter of putting the environment first but also about recognising that the economic viability of cities must built on a sustainable basis of long-term social, environmental and economic stability and equity. This issue goes to the heart of the mismatch of conceptions of quality of life, and the vital need to make fully clear the real costs of the pursuit of short-term quality of life at the expense of longer term sustainable development, and so to demonstrate that the shift

Box 1.2 Quality of life and sustainability

Undoubtedly, environmental and sustainability factors have great significance for quality of life, even if people are not always aware of it in daily life. Illustrating this point, Brundtland's definition of sustainability, the definition of sustainable development most commonly referred to, begins with human needs: 'Sustainable development meets the needs of the present generation without compromising the ability of future generations to meet their own needs.', and the World Commission on Environment and Development (WECD) further defines sustainable development as: 'A global process development that minimizes the use of environmental resources and reduces the impact on environmental sinks using processes that simultaneously improve economy and the quality of life.' Here, 'sustainability is the continuation of the quality of life for generations to come also including the proper distribution of quality of life between groups and with other parts of the world' (WCED, 1987).

The concept of sustainable development emphasizes the maintenance of natural resources and the natural environment as a prerequisite for developing any economic activity to achieve human well-being and quality of life. Nature provides 'life support mechanisms and services' as a basis for society. Economic activities are the means to utilize these resources and to release their potential value to society in order to meet human needs. According to this model, economizing is the human activity that continually converts natural resources into quality of life as expressed in terms of goods and services (Figure 1.1). Clearly, a healthy environment and the wise use of natural resources are indispensable for sustainable development which provides the basis for long-term quality of life.

Figure 1.1 What should be sustainably managed?

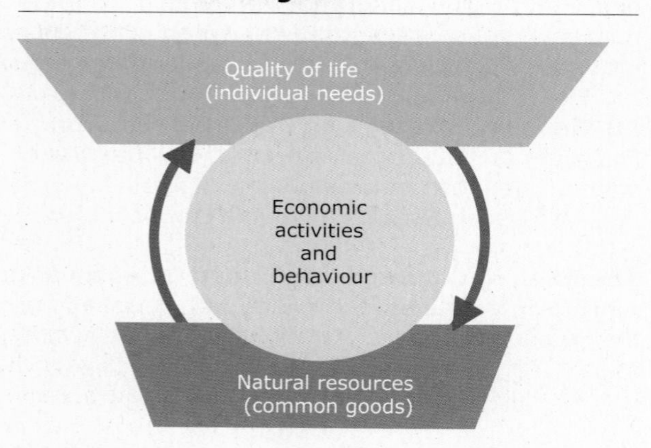

Source: ICLEI — Local Governments for Sustainability, 2008.

to a more sustainable way of life does not involve a loss but rather a real increase in quality of life for all population groups as well as future generations. This clarity can then provide support for the policy-makers and the means for cities and towns to deliver more sustainable ways to achieve quality of life.

1.2 Quality of life — visions or preferences?

Almost three quarters of European citizens live in urban areas today, and this is expected to increase to 80 % by 2020. In many respects the European Union can be seen as a Union of cities, as approximately 1 600 urban areas with more than 50 000 inhabitants are defined as functional urban areas (Box 1.3) (ESPON, 2005a).

According to the results of a survey of 75 cities across Europe (EC, 2007a), the overwhelming majority of citizens seem satisfied with their quality of life in the city. The precise alchemy of quality of life in a town or city remains obscure, apparently defying definition. One definition is that quality of life exists when people can live a healthy, pleasant and safe life, 'can be who they want to be and do what they want to do' (Sen, 2003). But individuals have their own visions and preferences, leading to a great diversity of personal definitions, as some examples from France demonstrate (Box 1.4).

Nevertheless, the basic idea of what constitutes quality of life is much the same throughout Europe. There are common concerns for all, including making a living and having an income, enjoying a satisfactory family life, and having good health (see also Box 1.5). Although at the individual level, assessments of the importance of these factors vary (Eurofound, 2004 and 2007).

Urban dwellers have only subtly different views on quality of life. Asked to give a definition of

Box 1.3 Definitions of urban areas

Urban areas can be defined according to different criteria. Apart from population thresholds these include:

Administrative area: constitutes the territorial expression of the political and technical framework of governance, forming the focus for, and critical to the understanding of, the development and implementation of policies to secure both quality of life and sustainable development.

Morphological area: constitutes, irrespective of administrative borders, the spatial dimension and form of cities and towns in physical terms, comprising urban fabric with buildings, roads and artificially surfaced area, industrial and commercial units, green urban areas within urban fabric, and in addition port areas, airports, and sport and leisure facilities if included or continuous to other urban land use.

Photo: © Image 2000

Functional urban area: constitutes the socio-economic reality of town and cities expressed in terms of the territorial influence of the town or city across its hinterland, and identified in the relevant structures of the built environment. The functional urban area normally embodies contrasting town — suburb and rural areas and forms the focus for the socio-economic and environmental forces that mould the development of towns and cities. These forces include, for example, the decentralising trends within the functional urban area that generate the intra-regional migration patterns.

Using different delineations: the relationships between administrative, morphological and the functional urban areas and their appropriate consideration by decision-makers is critical to the effective governance of the forces creating the social, economic and environmental challenges posed by towns and cities. Due to inertia in the re-definition of administrative areas, the functional urban area typically extends well beyond the administrative and morphological boundaries of the town or city. These considerations, which concern the relationships between the socio-economic driving forces of change and the administrative units of governance designed to manage change, highlight the need to secure both vertical and horizontal policy integration for the effective governance of towns and cities.

their own quality of life apart from income, most people emphasise public services, employment, shopping, transport, green open space, culture and sporting facilities, as well as space to live. All of these factors make a city attractive. When people are further asked 'What could be done to improve the quality of life in your town?', they tend to mention problems they face on a daily basis, such as traffic, noise and air pollution. Environment is seldom mentioned directly as poor environment is mainly seen as a price to they have to pay for the advantages of the big city. Nonetheless, the search for a better environment is a reason people give for moving out of the city itself while staying close enough to benefit from urban facilities. The individual search for a better quality of life is therefore a matter of trade offs: paying less for a bigger house, but spending more time in traffic jams or sacrificing urban amenities for a better environment.

In conclusion, the dimensions of quality of life are diverse, and some factors are undeniably more important drivers of change in towns and cities than others, but these diverse dimensions and drivers are always interrelated.

Definitions of quality of life?

Another aspect to consider when describing quality of life is that it has an objective and a subjective perspective. The concept of quality of

Box 1.4 French definitions of quality of life

In replying to the interview question: 'What comes first to your mind when I say quality of life in your city and region?' people gave different answers; among these:

- 'Quality of life... purity... environment also. Nuisances, no nuisances, the calm ... green, birds' (Woman, 40 years old, Paris region);
- 'This would be about living in a nice green, small town, like Chaville (Paris region). Well connected with transport, by bus or train. Not so far from Paris, but still far enough. We are near to forests, which is good for having a good air, and we have all the shops, markets, public services, everything at hand' (Woman, 55, Paris region);
- '... having good human relationships' (Man, 35, Paris);
- 'It means money, to have enough money for a living, also the surroundings, the environment where we live' (Woman, 28, Vélizy, Paris region);
- 'I live in Marseille, that's why. I wish I lived outside of the town, I prefer nature. The traffic, the noise. To go for a walk, to go to the sea shore, it takes 30 minutes by car, traffic jams included. In a small town, it would take five minutes to be in the nature' (Man, 27, Marseille);
- 'The most important: health, human environment, nature, that everything be respected, that pollution would stop, that we would take care of nature as we must. Beyond all, health — this is the most important' (Woman, 72, Nice);
- 'The relation to the working environment, even if I am now retired, seems important to me, also the leisure, and everything around social relations, more generally the relationship to the others' (Man, 68, Cachan, Paris region);
- 'It makes me think of the sun, this is very important for me. Further having a nice little garden, try to avoid big town pollution, try to have a hygienic life, avoid the stress of big towns, so having a nice little house at the countryside (his house is 120m^2!), not far from the commodities of the modern world, to be able to use them without being dominated by them, such as supermarkets, cinemas, restaurants or other leisure possibilities' (Man, 45, countryside 30 km outside of Nice).

Source: Results of research interviews lead by Michelle Dobré in 1999 in Paris, Nice, and their regions with the support of the DRIRE IdF and PACA.

Box 1.5 Domains of quality of life

The first Survey on Quality of Life in Europe 2003 investigated 8 domains of individual life situations in 25 Member States. These do not cover all aspects but the most relevant for a complete description of quality of life in both its objective and subjective dimensions.

- Economic situation;
- Housing and the local environment;
- Employment, education and skills;
- Household structure and family relations;
- Work-life balance;
- Health and health care;
- Subjective well-being;
- Perceived quality of society.

Source: Eurofound, 2004. www.eurofound.europa.eu/pubdocs/2004/105/en/1/ef04105en.pdf.

life was popularised from the early 1950s, and in the context of economic growth, quality of life referred to individual happiness and well-being. The concept emerged as a response to objective measures of material progress including gross domestic product (GDP); it provides indicators for other, material and non-material criteria and of subjective views on the human condition. Early studies on quality of life demonstrated that growth in objective material comfort was not necessarily matched by similar growth in satisfaction, well-being or happiness (Campbell *et al.*, 1976; Andrews & Withey, 1976) and therefore indicates the need to consider both perspectives.

The objective perspective highlights issues such as income level, living conditions, job situation. The subjective approach focuses on individual appreciation of these issues (Box 1.5); for example, 40 m^2 of living space per person might be perceived as luxury in one country and seen as only standard in another country. From an urban planning perspective, quality of place (Massam, 2002) describes the state of the external environment and the requirements for good quality of life. This approach to quality of life deploys various socio-economic and environmental indicators, such as air or water quality and material welfare.

City rankings

To aid understanding of the underlying reasons for differences, and thereby support policy definition and implementation, there have been many attempts, some more scientific than others, to rank cities in terms of quality of life. In June 2008 the *Copenhagen Post* reported proudly 'Copenhagen best city to live'. This was in response to an article in the UK magazine *Monocle,* which ranked Copenhagen as the best city in the world to live for quality and design compared to 25 other cities. However, the newspaper also commented that whilst Copenhagen is clearly a great city, '...even the city's most enthusiastic residents should take it ... with a grain of salt... Anyone who lives here for more than a week will tell you that its recent ranking as the 'world's best city for quality of life' is absurd.' This one example illustrates the difficulties with quality of life ranking of towns and cities. Reviewing the various city rankings from *Mercer, Readers Digest* and many others reveals that the rankings can differ widely according to the index criteria. While the rankings highlight useful similarities and differences, the evident contradictions also question the validity of simple comparisons.

Citizens, local authorities, politicians and businesses are all sensitive to city rankings, even though it is commonly acknowledged that ranking is virtually impossible, providing at best only a partial picture of reality. Rankings can be biased and/or contradictory, dependent on the ranking criteria. City rankings as communication tools reflect a desire to simplify complexity and to guide action. Consequently, they can be useful tools for policy-makers but must be viewed in context.

1.3 Health, environment and social equity: basic quality of life indicators

Together with growing incomes, better paid jobs and rising levels of education, good health and secure family and social relations remain key determinants of individual happiness and fulfilment (Eurofound, 2008). The urban environment influences human physical, social and mental well-being, therefore, a healthy, supportive environment is indispensable to quality of life in cities. People need to breathe clean air, have access to clean drinking water and adequate housing conditions, and enjoy quiet and peaceful places. Accessible, good-quality, well-maintained green spaces and playgrounds, modern transport systems and safe, walkable neighbourhoods that encourage physical activity and social interactions are key constituents of urban quality of life.

Urban design and planning

Characteristics such as population density and the extent of sealed areas are comparable for and define urban areas. Such areas differ from the rural environment and generate, for example, the urban heat island effect. However, the actual impact on the urban environment is dependent on specific local characteristics, which differ from city to city.

Well-designed buildings and public spaces in a well-planned urban environment can provide attractive, secure, quiet, clean, energy-efficient and durable surroundings, in which prosperous and healthy communities can thrive in the long term. The World Health Organization (WHO) considers urban planning an important determinant of health, and also economic development — as the attractiveness of a city or town is becoming an increasingly important factor in the decision-making process. However, the realisation of a healthy urban environment in which all determinants of healthy living are integrated in a holistic manner is a challenging objective, as Chapter 2 of this report will demonstrate. Urban design and building regulations are both very important in this respect.

The following paragraphs provide some illustrations of health, environment, social equity and urban design features, from both objective and individual perspectives as basic elements of individual quality of life. However, these illustrations are only partial as the impacts of the urban environment on health and quality of life are not distributed equally; frequently, children, the elderly and those living in deprived urban neighbourhoods are disadvantaged.

Social equity and housing conditions

Environmental and health impacts are not equally distributed throughout Europe or within cities. In the United Kingdom in 2004, 20 % of those in the lowest income groups lived in poor quality environments compared to 11 % of those in the highest income groups (1) (UK Office for National Statistics, 2007) emphasising the fact that inequalities in quality of life reflect inequalities in economic, social and living conditions. Poorer people, immigrants, and other disadvantaged groups typically inhabit the worst parts of the city, for example near contaminated sites, and are more affected by the lack of green space and public transport services, by noisy and dirty roads and by industrial pollution.

Perceived safety and the socio-economic status of an area seems to play a key part in determining urban quality of life and also influences physical activity, obesity and related health problems. Studies in eight European cities found that residents in areas with high levels of graffiti, litter and dog mess were 50 % less likely to be physically active and twice as likely to be overweight (Sustainable Development Commission UK, 2008). Furthermore, the 2003 Health Survey for England suggests that perceptions of social disturbance in neighbourhoods are associated with higher risks of obesity and poor health, whereas positive perceptions of the social environment have the opposite association. Areas with a high socio-economic status tend to have better quality recreational environments when compared to low status areas, and people who live in high status areas tend to be more active in leisure time (Kavanagh *et al.*, 2005). Accordingly, feeling safe in the neighbourhood is likely to increase levels of physical activity. Natural features, especially in underprivileged neighbourhoods, can encourage people to walk, cycle and play outdoors and socialise, so facilitating social integration.

Adequate housing conditions are also important determinants of quality of life. People living in low standard buildings with poor energy performance and in 'fuel poverty' (2) experience problems with both excessive cold and heat. Cold is a major cause of winter death, particularly amongst the elderly. Cold, poor ventilation and inadequate heating contribute to dampness and consequent health problems. Poor indoor air quality, poor construction, poor maintenance of housing and individual lifestyles all influence residents' health.

Impacts of air pollution

The EU estimates that human exposure to fine particulate matter ($PM_{2.5}$) (3) causes about 350 000 premature deaths each year. In other words, at these exposure levels the average life expectancy is reduced by almost a year — almost two years in the most affected urban areas of Belgium, the Netherlands, Northern Italy and parts of Poland and Hungary (EEA, 2007b). The major air pollutants in urban areas are particulate matter, ozone and nitrogen oxides (NO_X). These pollutants pose serious threats to human health, as they can cause respiratory disorders, aggravate asthma, and impair development of lung function in children. Measurements of air quality show that almost 90 % of the inhabitants of European cities where PM_{10} concentrations are measured are exposed to concentrations that exceed the WHO air quality guideline level of 20 $\mu g/m^3$.

The overwhelming majority of people surveyed in 62 of the 75 European cities participating in the urban perception survey (EC, 2007a) agreed that air pollution is a major problem. Compared with the measured data on NO_2 and PM_{10}, these perceptions correspond closely with the objective situation.

(1) In the United Kingdom, a special multidimensional index of deprived areas is used to identify the 'critical locations in urban setting', based on information on employment, health, income, education and skills, barriers to services, crime and living environment, including air quality, distance from a waste disposal site, proportion of people living near the regulated industrial source, and proportion of people at significant risk of flooding.

(2) A household is in fuel poverty if it has to spend more than 10 % of total household income on energy in order to sustain comfortable conditions.

(3) $PM_{2.5}$ is particulate matter with an aerodynamic diameter of up to 2.5 µm and PM_{10} up to 10 µm The estimate is based on model calculations using anthropogenic primary PM and PM precursor emissions as an input (year 2000, EU-25) EU Clean Air for Europe (CAFÉ) programme http://europa.eu/scadplus/leg/en/lvb/l28026.htm.

Figure 1.2 Perceived and reported air pollution

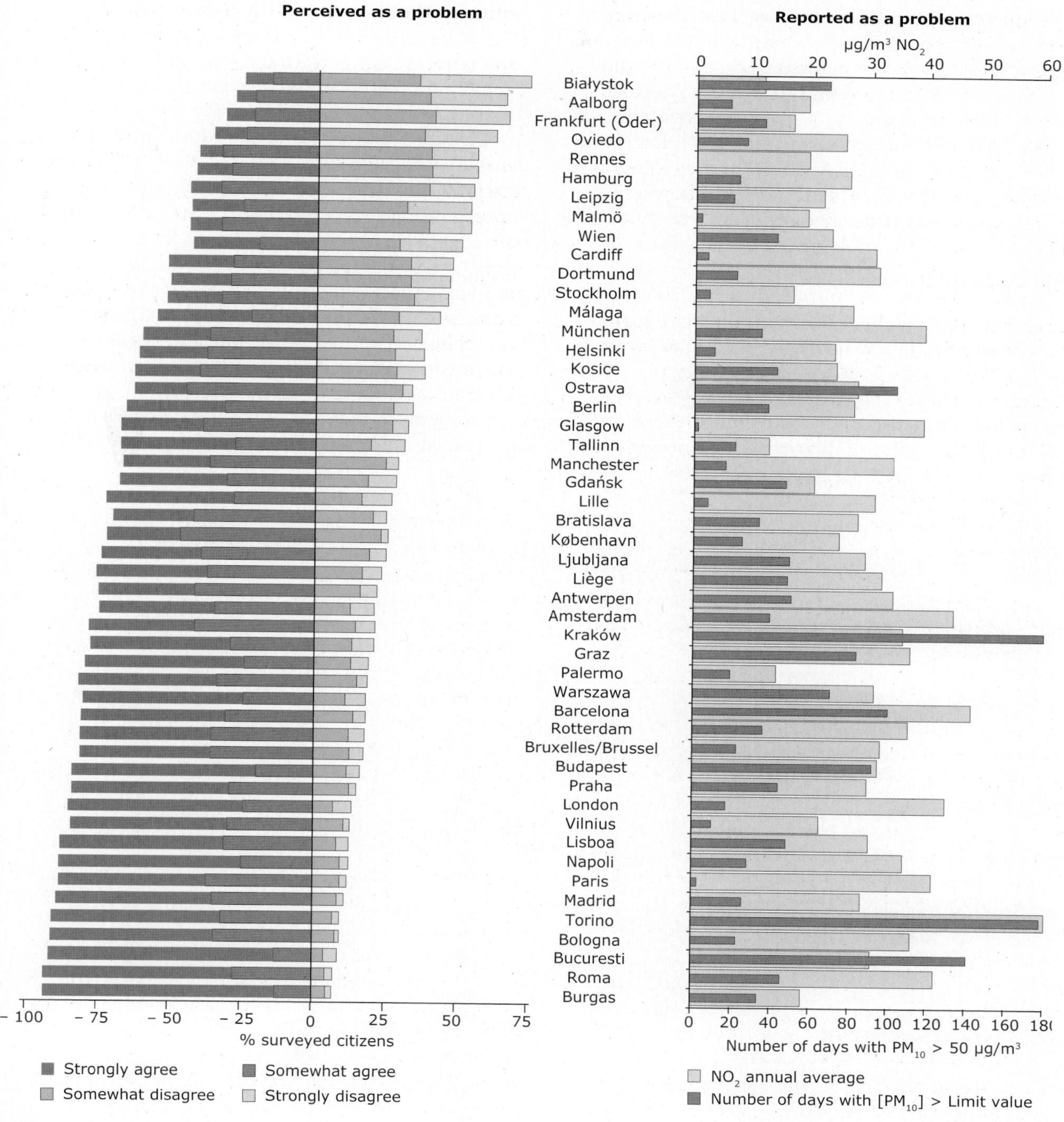

Source: EC, 2007a and AirBase.

However, a number of discrepancies (Figure 1.2) suggest that additional factors such as the general image of the city, its attractiveness, available green space or levels of noise also play a role and influence individual perceptions.

Impacts of noise exposure

Environmental noise affects health and urban quality of life by interfering with sleep rest, study and personal communication. Chronic exposure to noise is associated with increased risk of heart disease, hearing impairment and impacts on mental health. These effects may be enhanced by interaction with other environmental stressors, such as air pollution. For example, in Germany approximately 3 % of acute myocardial infarctions may be attributed to road traffic noise (Babisch, 2006). Also in Germany, 60 % of the population are adversely affected by road traffic noise, and 10 % are highly affected (UBA, 2005). In the Netherlands, 29 % of the participants in a national survey are troubled by road traffic noise, mostly from mopeds (RIVM, 2004). The most troublesome sources of noises are transport, primarily roads, railways and aircraft. Furthermore, noise problems are often worse in areas of high density housing, deprived neighbourhoods and in rented accommodation.

Figure 1.3 shows the variance of noise levels in some European cities. In some cities the majority of residents are living in areas with a noise level of more than 55 dB — a level associated with significant annoyance. However, as with air quality, Figure 1.3 shows that perception of noise can differ markedly from that reported. Furthermore, whilst the perception of noise as a problem is more or less the same in Malmö, Ostrava, Leipzig and Munich, in reality, a much larger percentage of people are affected by high-noise levels in Malmö and Ostrava than in Leipzig and Munich. Some of these apparent differences may, of course, be attributed to differences in noise modelling or survey methods.

Impacts of climate change

Climate change raises new, complex challenges for the urban quality of life and the health of European citizens. High population densities means that cities are highly aware and concerned of problems associated with climate change. Cities rely on complex systems to deliver power, water,

Figure 1.3 Perceived and reported noise pollution

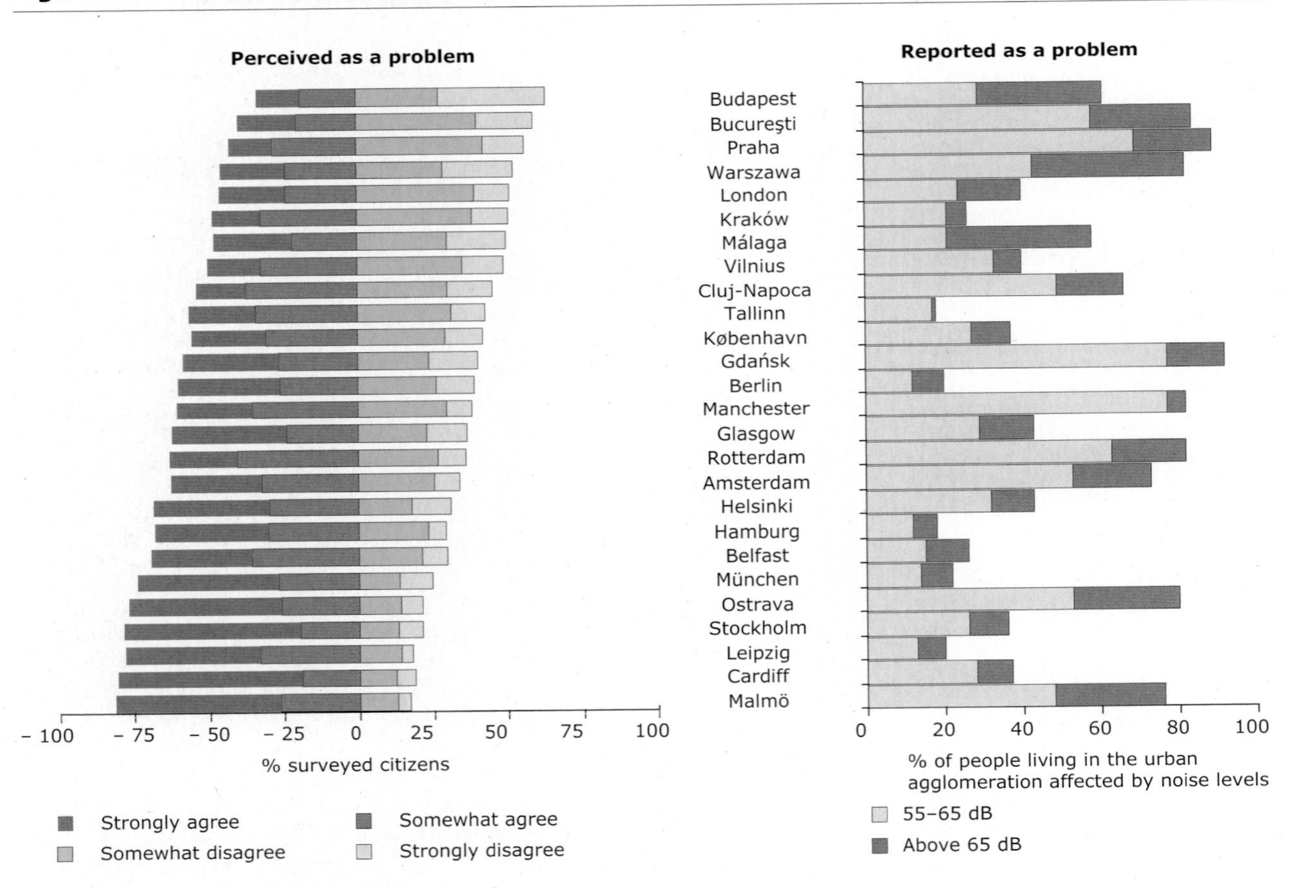

Source: EC, 2007a and 2007 data reported under the Directive on environmental noise (EEA, 2008). Note that the noise exposure data are that which has been reported by Member States in accordance with the END until 31 October 2008. At the time of writing, some of this data may not have been subject to a full quality assurance check.

Map 1.1 Number of tropical nights over Europe

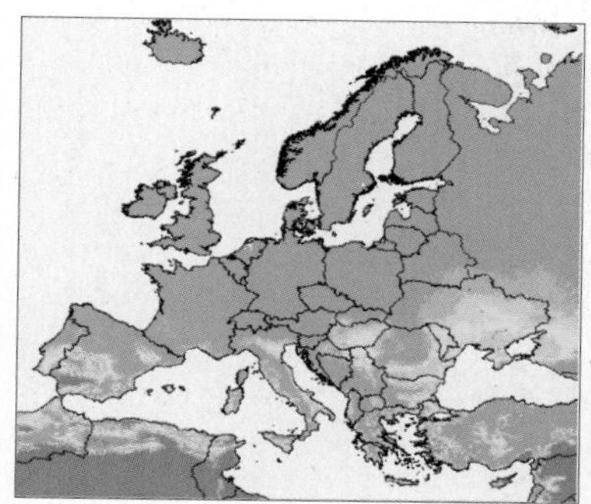

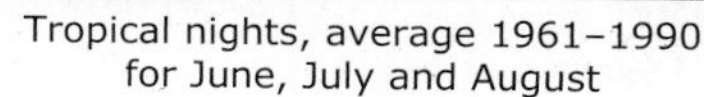

Tropical nights, average 1961–1990 for June, July and August

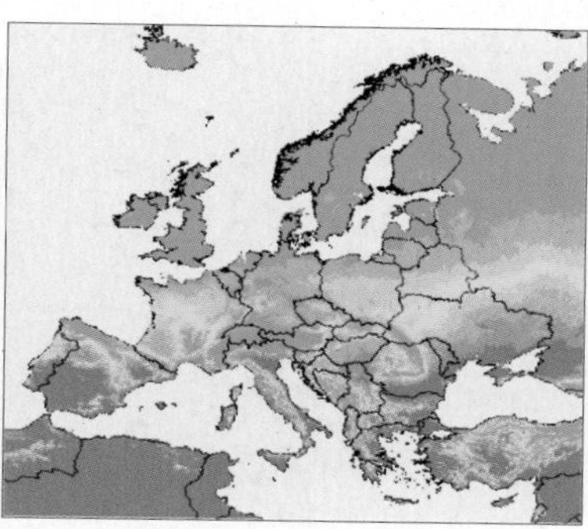

Tropical nights, average 2071–2100 for June, July and August

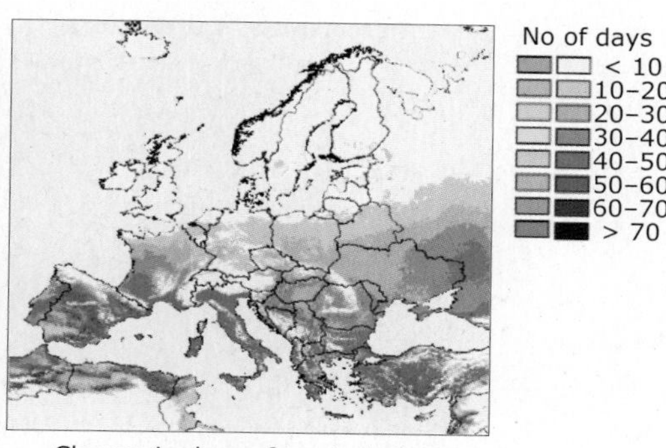

Change in days of tropical nights from control to scenario period for June, July and August

Note: Number of tropical nights (i.e. minimum temperature > 20° C) over Europe for the control period (1961–1990) and the scenario period (2071–2100) during summer seasons (June–August) and change between periods (right panel).

Source: Dankers and Hiederer, 2008.

communications, transport and waste disposal, and soil sealing increases the risk of flooding, drinking water shortage and the spread of infectious disease. Health impacts of heat waves are more pronounced for vulnerable groups, such as infants, children, the elderly, and those living in deprived areas and are unable to take remedial action. More extreme weather events including floods, droughts, and heat waves are already more evident throughout Europe: it has been estimated that the 2003 heat wave caused more than 52 000 premature deaths (EPI, 2006).

Green urban spaces

Studies in the Netherlands demonstrate that children with good access to green open space, fewer high-rise buildings and more outdoor sports facilities are more physically active. Similarly, studies of eight European cities show that people who live in areas with abundant green open space are three times more likely to be physically active and 40 % cent less likely to be overweight or obese (Ellaway *et al.*, 2005). School children who have access to, or even sight of, the natural environment show higher levels of attention than those without these benefits (Velarde *et al.*, 2007).

Green areas are important for health because they:

- allow for contact with nature, promote recovery from stress, are beneficial for mental health and help improve behaviour and attention in children;
- improve air quality and help reduce heat stress;
- encourage people to be physically active.

The urban perception survey (EC, 2007a) demonstrated that the majority of respondents in Northern European cities were satisfied with the supply and quality of green areas. However, there can be large differences between the perceptions and the actual proportion of the urban area devoted to green open space. For example, in the municipality of Brussels, where there are few areas of green space, most respondents expressed satisfaction with the supply of green open space; whereas in Bratislava, where there are large areas of green open space, the level of satisfaction was much lower (Figure 1.4). Some of the discrepancies may be the result of statistical sampling effects and cultural differences; however, these results do seem to indicate that it is not only the total area that is important in individual satisfaction, but also the quality of green open space, including accessibility, possibilities for outdoor recreation, distribution and the overall design of the urban area.

Space for pedestrians and cyclists

Good quality, accessible and safe walkable neighbourhoods encourage daily physical activity such as walking and cycling. These factors help combat the health impacts of sedentary lifestyles, especially in relation to obesity and cardiovascular disease. Public green open space provides opportunities for exercise. People are more likely to walk, cycle and play in natural spaces, enjoying the benefits of physical activity and social interaction. For example, in Maastricht in the Netherlands neighbourhoods with nearby sports facilities or parks are positively associated with time spent cycling (Wendel-Vos *et al.*, 2004).

Figure 1.4 Perceived and reported green space

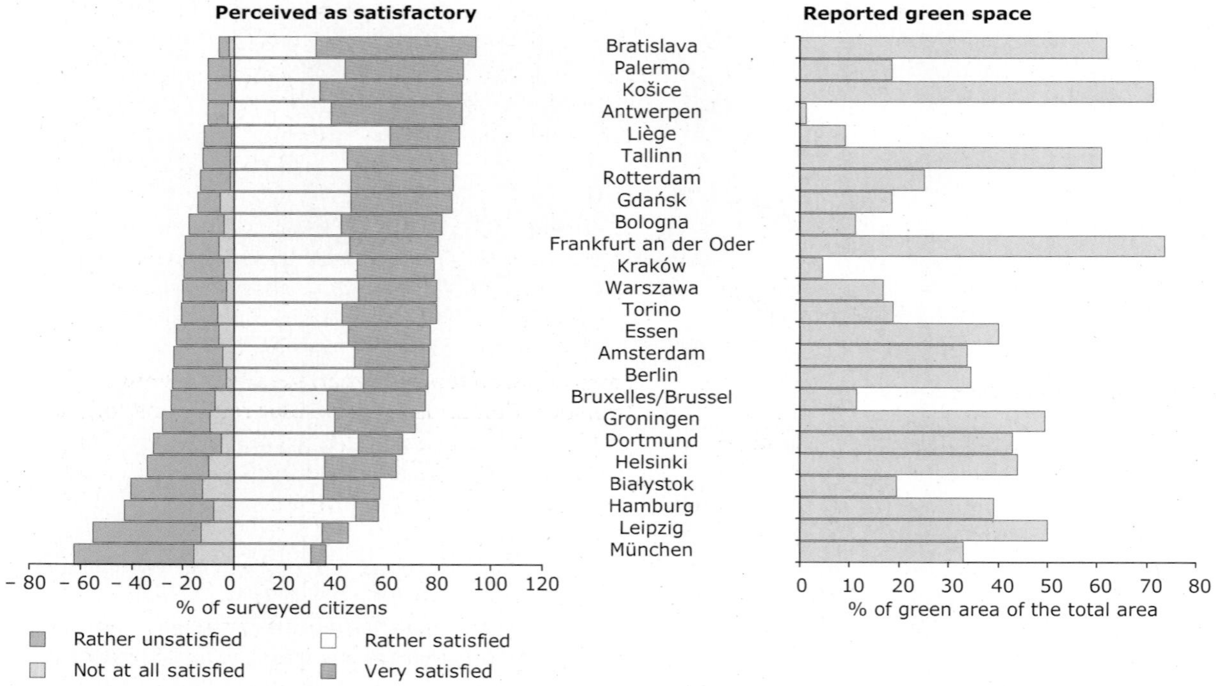

Perceived as satisfactory
Reported green space
Bratislava
Palermo
Košice
Antwerpen
Liège
Tallinn
Rotterdam
Gdańsk
Bologna
Frankfurt an der Oder
Kraków
Warszawa
Torino
Essen
Amsterdam
Berlin
Bruxelles/Brussel
Groningen
Dortmund
Helsinki
Białystok
Hamburg
Leipzig
München
- 80 - 60 - 40 - 20 0 20 40 60 80 100 120
% of surveyed citizens
0 10 20 30 40 50 60 70 80
% of green area of the total area
Rather unsatisfied
Not at all satisfied
Rather satisfied
Very satisfied

Source: EC, 2007a and Urban Audit Database, data 2004 on core cities (Eurostat).

The provision of cycling and pedestrian infrastructures is both quantitatively and qualitatively important. Figure 1.5 illustrates major differences throughout Europe leading to big differences in rates of both walking and cycling — the latter ranging from below 1 % of people cycling to work to around 36 % in Copenhagen. The quality of transport infrastructure has a major influence on walking and cycling in cities, but it does not explain all differences. Other factors such as city structure, safety, geography and cultural

Figure 1.5 Cycling paths and lanes in European cities

Cycling paths and lanes per city area in km/km²
8.9
0
1
2
3
4
5
6
Aalborg
Aarhus
Barcelona
Berlin
Bristol
Brussels
Copenhagen
Dresden
Goteborg
Hannover
Heidelberg
Helsinki
London
G.Lyon
Madrid
Milan
Munich
Naples
Paris
Patras
Prague
Riga
Rome
Stockholm
Tampere
Turku
Wien
Oslo
Zaragoza

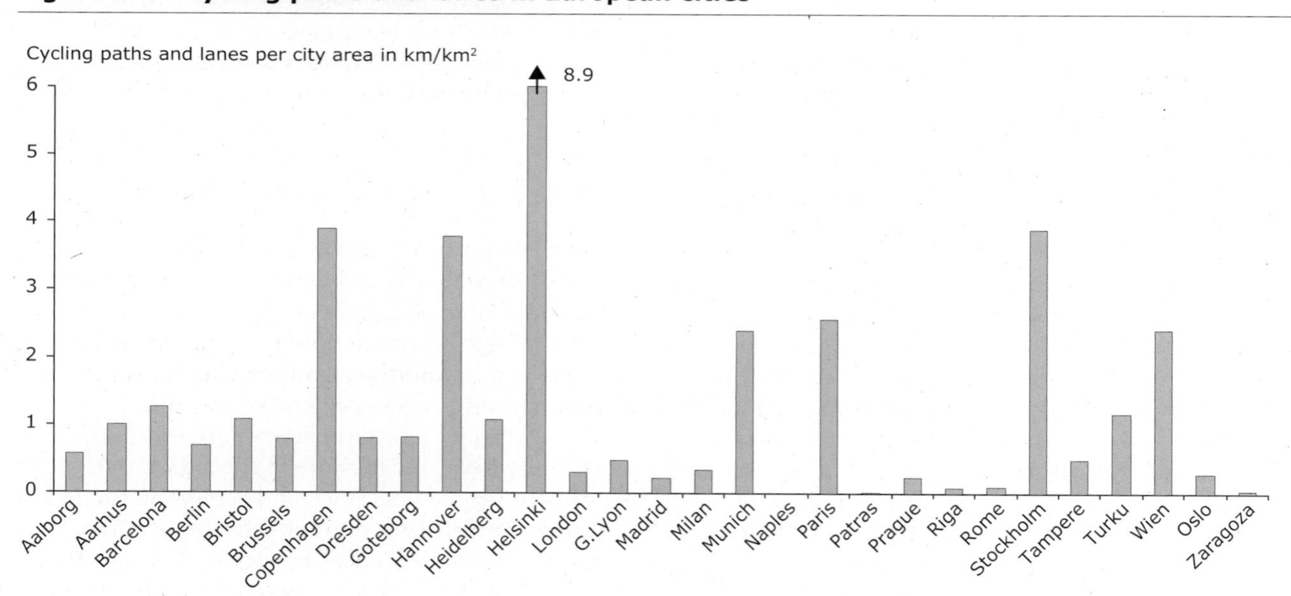

Source: Ambiente Italia, 2007.

needs should be considered as well. Fundamentally, as explored in the next chapters, different local responses can be explained by different conceptions of quality of life, leading policy-makers to diverging recommendations on what should be done in order to improve quality of life in Europe's cities and regions.

1.4 Cities and towns determine Europe's quality of life

The fight to tackle climate change will be won or lost in cities, Ken Livingstone, Mayor of London, 2007.

The potential of cities is there…

Growing cities and changing lifestyles demand an ever-increasing supply of natural resources. Cities occupy just 2 % of the world's surface, but at the same time, are home to half of the world's population, which is responsible for three quarters of natural resources consumed globally (UNEP, 2008). Cities are hugely reliant on regions and nations well beyond their own boundaries and have many interactions with local and global hinterlands. A city depends on resources produced outside the city and transported to the city for consumption, and the waste products of consumption in the city are disposed of elsewhere. Consequently, cycles of production and consumption and their environmental impacts cannot be separated. Europe is already highly urbanised, and cities and towns, by virtue of these relations with their hinterlands, substantially determine the potential for sustainable development and quality of life for both urban and rural areas.

…but cities perform differently!

The ecological footprint of a city provides a means of assessing how much land and water any individual city virtually requires to produce the services and resources it needs and to absorb the waste generated. The footprint is normally expressed in terms of spatial extention of land and water from the city, which in the case of London extends to over twice the area of the United Kingdom. This indicator can reveal differences in performance, prompting questions about the underlying reasons and so stimulating further investigation of the causes and potentials for action. From a European perspective, cities' ecological footprint can raise awareness of their overall impact on the European environment.

The concentration of population, consequent levels of service provision and urban lifestyles mean the ecological footprint of cities is generally higher than that of rural areas of the same size. However, individual city dwellers tend to have a lower average ecological footprint than those living in rural areas. This is primarily because most city residents have shorter distances to travel to work, while many rural residents commute long distances to work, typically by car. Also, urban housing is normally more efficient in terms of energy consumption. As a result, urban lifestyles can offer the potential to lower the overall regional or national footprint and environmental impact. These conclusions are critical to the arguments that cities and towns offer the best hope for living more sustainably and reinforce the argument for compact cities, as clearly demonstrated by the London transport ecological footprint (Box 1.6). Conversely, urban sprawl, growing transport demands, in particular road transport, as well as current urban lifestyle choices demanding goods and services from a global hinterland, tend to increase the ecological footprint of cities.

Revitalising actions by cities and towns

It is clear from the above that the nature of cities substantially influences the quality of both urban and rural life, and that quality of life can be enhanced by improving the way cities are managed. City mangers have the power to drive forward change and reduce the negative impacts of urban development such as urban sprawl and growing demands for car-based urban transport. They can do this by developing and implementing policies for urban planning, urban design, housing and local transport, thereby offering new opportunities for more sustainable lifestyles and quality of life. Urban planning and urban design are fundamentally local responsibilities. City and regional planning guides the functional organisation of the city, which in turn sets the framework for the patterns of urban consumption and the basis for realisation of quality of life in cities. The compact city based on efficient public transport, provision for walking and cycling allied with high quality public and green open spaces can provide the model for enhanced quality of life and sustainable development.

Cities are also the focus of the consumption of energy and other resources. Cities can therefore act decisively to combat resource depletion and mitigate climate change by, for example, avoiding energy-intensive transport and promoting energy-saving housing policies, as well as containing urban sprawl.

Local authorities have the legal power, and responsibility to regulate and manage urban policy and implement effective planning strategies in the interests of their population. However, it is clear that cities cannot be managed in isolation from the many powerful forces and decisions originating outside

Box 1.6 London transport ecological footprint

Londoners have the second-lowest transport footprint per resident out of 60 British cities despite coming 44th in the list of overall footprints. This is because London has a good, well-used public transport system at affordable prices. London is at an advantage because it has such a large number of people concentrated in a small area, which makes running public transport a more attractive proposition. There are also disincentives to car ownership in the city, such as limited car-parking and the congestion charge for central London. In contrast the Home Counties have very large transport footprints. This is most likely due to people commuting into London, as well as the relative level of affluence (which is related to higher levels of car ownership) in the Home Counties.

Photo: © Jens Georgi

Figure 1.6 UK city transport ecological footprints

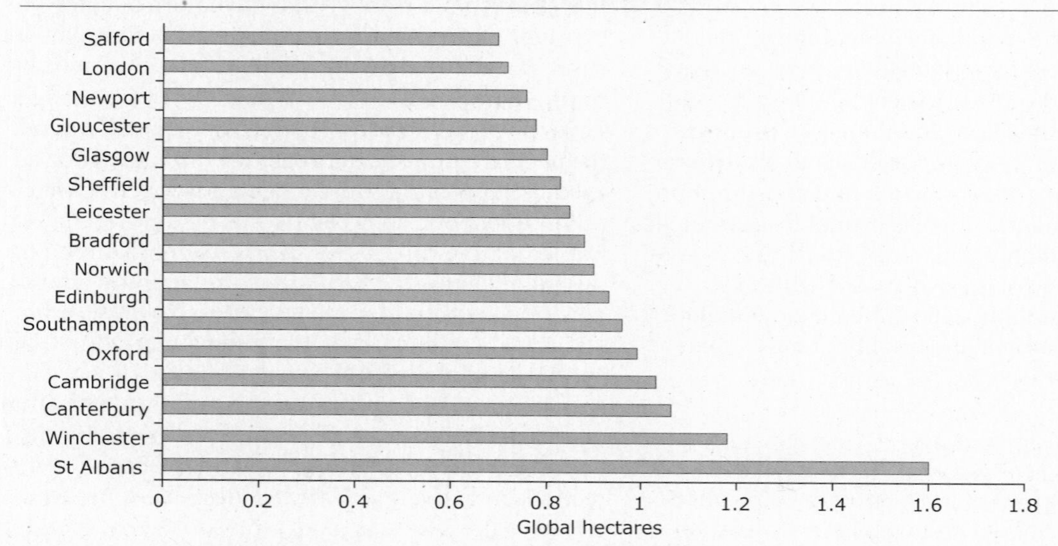

Note: Selection from 60 cities including the cities with the lowest and highest transport footprint. See more in the study mentioned as source.

Source: WWF-UK, 2007, http://www.wwf.org.uk/oneplanet/cf_0000004481.asp.

their boundaries. Local policies must therefore be complemented by regional, national and European policies to effectively address these current and future challenges as outlined in the next section.

1.5 EU and urban policies interact

As described in Section 1.4, cities are no longer isolated and self-sustaining units but are strongly linked via their functional urban areas with the towns and settlements of their hinterland, as well as with other cities in their region, in Europe and globally. Cities are therefore subject to many European and global challenges driven by forces outside their direct control, which they must respond to in order to ensure quality of life for their inhabitants.

The major drivers of these European and global challenges include the new potentials of information technology, which are rapidly transforming the accessibility of cities, as well as major demographic changes, including the general aging of the European population and continued migration. Cities and towns need to respond to the economic, social and environmental consequences of the individual highly material lifestyles of their citizens fostered by a political climate in which growth in welfare and enhanced quality of life is still equated with growth in GDP (see also Section 2.2). Overall, cities and towns find themselves in an extremely complex situation.

European policy — a clear role

European policy combined with policy initiatives at the local level have the potential to drive and direct these major trends towards an enhanced quality of life in European cities. European climate change policy, for example, aims to mitigate the effects of climate change on urban areas. European cohesion policy supports the EU Lisbon Strategy for growth and jobs, and aims to improve the economic attractiveness of European regions. Together with other policies at various levels, the Lisbon Strategy aims to strengthen the economic basis of the regions and cities of Europe (Table 1.2). In support of the Lisbon Strategy the *Green Paper on territorial cohesion* (EC, 2008c) aims to transform territorial diversity into a key driving force for sustainable development.

Certain European policies, such as the EU Directives on ambient air quality and on environmental noise, also address the urban level directly. Other EU policies provide guidance for cities, including the *Community strategic guidelines of cohesion policy 2007–2013* (EC, 2006b), the *Thematic strategy on the urban environment* (EC, 2006d) and the *Leipzig Charter on sustainable European cities*.

Typically, European policy influences the urban level indirectly, and aims to support positive developments at the local level. However, due to a variety of factors, including inadequate policy coordination, there remains the risk, that policy implementation at the local level is in fact undermined by policy initiatives at EU level. Some of these effects are illustrated in Box 1.17 and Table 1.2.

More and better concerted action is key

Cities are now demonstrating an increasing understanding of the significant roles that they can perform in not only fulfilling EU regulations, but also in wider engagement in initiatives to secure sustainability in urban areas. These wider engagements include participation in the Local Agenda 21 processes, support for the Aalborg Commitments and the development of the Healthy Cities Network. These initiatives have provided a number of positive outcomes, including guidelines for policy development, as well as the exchange of good practice experience — although direct influence upon the evolution of EU urban policy has so far been limited. The ethos of these developments is expressed by the *Leipzig Charter on sustainable European cities* as follows: 'We must stop looking at urban development policy issues and decisions at the level of each city in isolation'.

Nonetheless, there remains much to do in fulfilling these objectives. Despite a growing awareness of the contributions that cities can make to the realisation of sustainable development, in many cases, cities still remain in relative isolation in the development of policy at the local level, apparently unaware of the need for a positive European dimension in city action. The *Thematic Strategy on the urban environment* (EC, 2006d) offers direct guidance on the sustainable management of cities, but does not explicitly require the development of integrated policy approaches that are linked to those at European level. However, the implementation of the Action Programme for the Territorial Agenda, and the follow up to the Leipzig Charter at local and member state levels offers real possibility for a new recognition of the need to adopt a more active and integrated approach to urban governance.

Box 1.7 European policy — positive and negative local impacts

Cohesion policy aims to support and strengthen cities and regions. Stronger cities and regions will provide their citizens with higher incomes increasing their material quality of life. At the same time this leads to changes in life style: Cars are more and more available and more used. People spend more time in leisure activities and make more vacation trips per year etc. with likely unintended negative environmental effects. Economically successful European and national funding which leads to stronger cities can also contribute to unbalanced price developments in particular for land, encouraging urban sprawl.

It is clear that cities exchange information and share best practice to support local sustainability, but joint and concerted action remains more the exception. A new approach adopted by the initiative of the Covenant of Mayors, in which major cities commit to reduce their CO_2 emissions by 20 % by 2020, may offer some further lessons on how to realise these potentials (Box 1.8).

The scenario of cities acting in isolation neglects not only the potential for concerted action, but also compounds the negative impacts of the common

Box 1.8 Covenant of Mayors act on climate change

On 29 January 2008, Commissioner Piebalgs launched the Covenant of Mayors, the most ambitious initiative of the European Commission involving cities and citizens in the fight against global warming. The Covenant of Mayors will be a result-oriented initiative in which participating local and regional authorities will formally commit to reduce their CO_2 emission by more than 20 % by 2020. In order to do that, they will develop and implement Sustainable Energy Action Plans and communicate on measures and actions taken their local stakeholders.

More information: http://ec.europa.eu/energy/climate_actions/mayors/index_en.htm.

Box 1.9 German Länder — competing municipalities

After German reunification in the early 1990's, East German municipalities prepared a large number of sites for commercial real estate development to attract major investments and jobs, with European and national funding. There was also a race between municipalities to obtain the highest share of the cake, but against expectations, no relation was found between the numbers of commercial real estate areas and economic development. Instead, the areas provided exceeded demand by three to four times. Today 30–40 % of the areas still lie idle and the remainder can easily satisfy the demand over the coming decades and even, sometimes, the next 100 years (BBR, 2005).

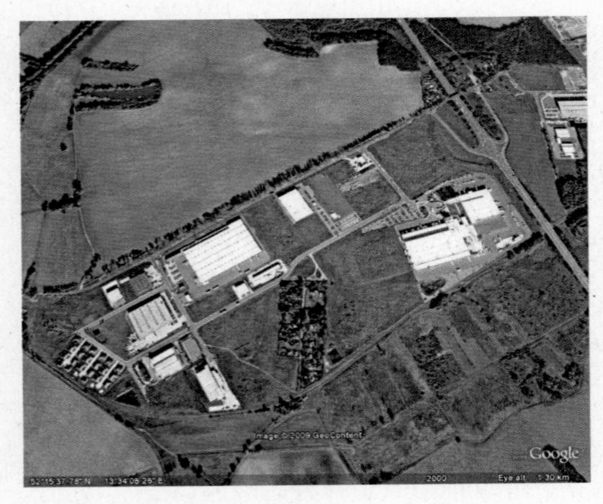

Photo: © Google Earth

Table 1.1 Länder of East Germany — commercial real estate capacity

	Commercial real estate area (brutto) in ha	Available commercial real estate area in ha
Mecklenburg-Western Pomeranian	9 785	4 070
Brandenburg	9 227	3 550
Saxony-Anhalt	4 500	1 840
Saxony	16 130	6 450
East Germany	49 570	1 920

Instead of improving the quality of life of their inhabitants by creating new jobs, many municipalities and funding authorities have wasted resources and even caused negative effects in the form of declining image and attractiveness, loss of biodiversity and ecological services — burdens that now have to be carried by the whole society.

Source: BBR, 2005.

tendency for cities to compete for limited economic resources, including industrial investments and national or EU funding. Competition tends to drive negative outcomes and unbalanced development, a game in which some individual cities win, but typically the net result is unsustainable development and lower quality of life for all (Box 1.9).

These negative impacts can be avoided by cooperation between cities in a regional policy framework that supports a holistic approach, integrating all agencies and government levels, as with the example of Berlin in cooperation with the neighbouring municipalities in Brandenburg (Box 1.10).

Integration: time to walk the talk

European, national and city policies can have a major impact on the quality of life in cities and towns (Table 1.2) demonstrating that cities and towns are not simply at the mercy of external drivers and processes. However, to date these potentials are not yet fully reflected in management and governance practice at the local level. Chapter 2 of this report further explores these issues in relation to specific challenges, and Chapter 3 provides some ideas and examples of the integrated approach to policy formulation and implementation that aim to fill the gaps evident today, and so provide the basis for the realisation of improved quality of life in the longer run.

Box 1.10 City-hinterland — cooperation Berlin and Brandenburg

From competition to inter-municipality cooperation

Initial situation

The fall of the Berlin wall and the reunification of Germany at the beginning of the 1990s led to massive suburbanisation over a wide area. This led to numerous conflicts of interest between Berlin and the neighbouring municipalities.

Solution

In 1996 the outer Berlin sub-districts and the neighbouring districts and municipalities in Brandenburg established the Municipal Neighbourhood Forum Berlin — Brandenburg. A series of working groups, presided over by the mayors and other stakeholders such as private companies and NGO's, provided a form for information exchange and discussion of spatial planning and development questions in the area. A secretariat of the Berlin government with its own budget supports the different activities, although the municipalities implement and fund them locally.

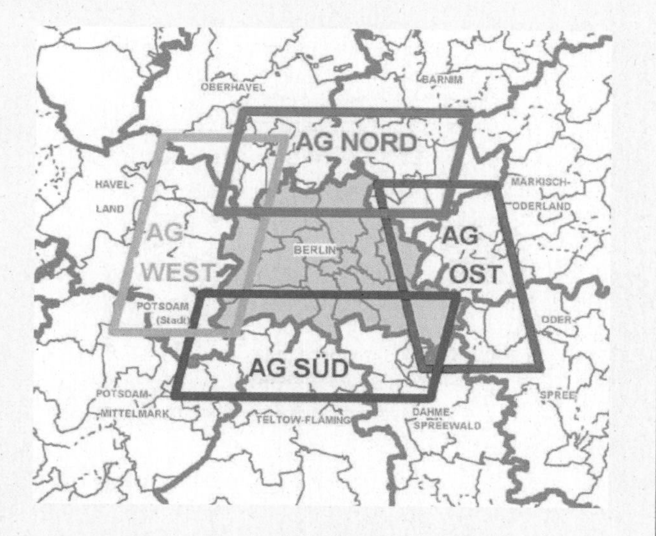

Source: Kommunales Nachbarschaftsforum, 2008.

Results

Long-term cooperation between the different stakeholders at the same level promoted the growth of understanding and awareness of spatial and cross-border interrelations, and supported the creation of joint responsibility for a balanced development of the area. Up to now, the partners have developed 11 structural concepts for different sub-areas, created the concept of a common bicycle route, and analysed the suburbanisation processes since 1990 in order to be prepared for future challenges such as demographic change. Generally speaking, the development has achieved a more rational and balanced basis than 15 years ago. Long term experience of transparent processes and mutual credit are the key factors for success.

Note: More information: http://kommunalesnachbarschaftsforum.berlin-brandenburg.de/.

Source: BMVBS, 2006.

Table 1.2 Major EU policy areas related to the urban level

Policy areas	Regional policy/ cohesion	Transport and energy	Enterprise	Environment	Employment and social affairs	Health and consumer protection	Agriculture and rural development	Research
	Lisbon Strategy and Gothenburg Strategy							
Aim	Improving the attractiveness of the European regions, to encourage innovation, entrepreneurship, growth of knowledge and to create more and better jobs.	Providing European citizens and businesses with competitive energy and transport systems and services.	Making the EU the most competitive and dynamic knowledge-driven economy.	Ensuring a high level of environmental protection. Contributing to a high level of quality of life and social well-being for citizens	Contributing to the development of a modern, innovative and sustainable European social model with more and better jobs in an inclusive society based on equal opportunities.	Ensuring a high level of protection of consumers' health, safety and economic interests as well as of public health at European Union level.	Promoting a robust and competitive agricultural sector. Contributing to sustainable development of rural areas.	Contributing to attain the objectives of the other Community policies.
Commitments 2008 total in billion EUR	36.6	2.8	0.6	0.4	11.5	0.7	54.1	4.0
Strategies and major directives with a relation to urban issues	Community Strategic Guidelines 2007–2013 for the Structural Funds. Green paper on territorial cohesion (2008). Communication on Cohesion policy and cities: the urban contribution to growth and jobs in the region (2006)	Green paper on urban transport (2007). Directive on the energy performance of buildings 2002/91/EC. Strategy for the simplification of the regulatory environment (2005).	Communication: Putting knowledge into practice: A broad-based innovation strategy for the EU. Communication: Implementing the Community Lisbon Programme — More Research and Innovation — Investing for Growth and Employment: A Common Approach.	Thematic Strategy on the Urban Environment. Several directives on: • ambient air quality; • environmental noise; • the Water Framework Directive; • Urban Waste water treatment; • Waste Management; • Emission standards, etc.	Social Agenda 2005–2010. European Employment Strategy and Employment guidelines.	White paper: Together for Health: A Strategic Approach for the EU 2008–2013. European Environment and Health strategy. Consumer Policy Strategy 2007–2013.	Common Agricultural Policy	FP 7 — Seventh Framework Programme
Programmes, instruments and funding (commitments for 2008)	Structural Funds 2007–2013 (explicitly including urban development): Cohesion Fund (8.1 billion EUR) ERDF — European Regional Development Fund (27.5 billion EUR).	Trans-European Transport Networks (TEN) — Financial support 2008: 0.9 billion EUR). Intelligent Energy — Europe programme (0.07 billion EUR in 2008).	Multiannual Programme for Enterprise and Entrepreneurship. ETAP — Environment Technology Action Plan. Competitiveness and Innovation Framework Programme (0.1 billion EUR).	EAP-The Sixth Environment Action Programme 2002–2012. LIFE+ programme — Financial Instrument for the Environment — 2007 to 2013 (0.25 billion EUR).	Structural Funds 2007–2013: European Social Fund (11.1 billion EUR).	programme of Community action in the field of public health 2008–2013 (0.045 billion EUR). Action programme environment and health 2004–2010. Consumer Programme 2007–2013 (0.02 billion EUR)	EARDF 2007 to 2013 — Rural development (12.9 billion EUR). AGF — direct aid, market support (36.8 billion EUR).	The FP7 bundles all research-related EU initiatives together under a common roof playing a crucial role in reaching the goals of growth, competitiveness and employment.
Linked to other European policies	Support other policies: transport, energy economy, employment, environment	Regional policy (accessibility), energy, environment, health, enterprise	Regional policy, employment, environment	Basis of life — Links to nearly all policies	Regional policy, economy	Public health is basis of life — Links to nearly all policies	Regional policy, employment, environment, health and consumer protection	Supporting the other policies with new knowledge

Tables 1.2 Major EU policy areas related to the urban level (cont.)

Policy areas	Regional policy/ cohesion	Transport and energy	Enterprise	Environment	Employment and social affairs	Health and consumer protection	Agriculture and rural development	Research
Potential impacts on the urban situation	↓	↓	↓	↓	↓	↓	↓	↓
Improving quality of life	Stronger, attractive and competitive cities; Renewal of cities; Polycentric territorial development.	Better accessibility of cities by road, rail, air, ship. Setting the framework for better urban transport. Promotion of sustainable energy systems and equipment and their market penetration.	Support eco-innovation in cities. Improvement of energy efficiency.	Reduce climate change impacts and background pollution levels, ensuring a healthy environment at all. Setting a framework to support sustainable urban development e.g. sustainable consumption pattern, right prices etc.	Reduce the economic, social and territorial disparities and strengthen the economic and social cohesion. Support the free movement of workers.	Setting the framework for high level of protection of consumers' health, safety and economic interests as well as of public health.	Ensure different services of rural areas for cities: food supply, water availability, nature, recreation etc.	Support cities by providing necessary knowledge.
Possible unintended side effects	Can promote unsustainable Western-European life styles all over Europe. Higher transport demand, higher energy and material use, urban sprawl. Competition between cities, regions.	Can encourage longer commuting distances and urban sprawl leading to even more transport. Unbalanced implementation of TEN can increase the share of road transport in relation to other modes.	Can lead to development of new residential areas and infrastructures around the new industrial clusters.	Can hinder some specific unsustainable economic development or consumer behaviour.	Likely effects on e.g. the environment depend on the way how growth and employment promoted.	Can hinder the production and distribution of some (dangerous) products and services.	Rural development, low prices for agricultural land, better accessibility can stimulate longer commuting and urban sprawl redefining the interaction with the cities.	
	↑	↑	↑	↑	↑	↑	↑	↑
Urban contribution to the European situation	Cities are motor of European development.	Organising an efficient and environmentally friendly urban transport. Increasing energy efficiency in housing and urban transport.	Cities are motor of European development and the place where most business takes place.	Support with their own environmental performance a high environmental quality across Europe.	Provide the majority of jobs and other socioeconomic services.	Contribute by their own measures to high levels of public health.	Provide socioeconomic and cultural services for the rural areas. Market of agricultural products.	Provide the practical case for research.

Note: Budget numbers from the Official Journal of the European Union L 71, Volume 51, 14 March 2008.

Source: European Commission, http://ec.europa.eu, 2008.

2 Quality of life and drivers of change

Frequently, processes and policies outside the direct control of cities and towns drive and determine their quality of life. Individual municipalities may feel at the mercy of such processes, but given the fact that in Europe urban areas contain nearly 75 % of the population, it is clear that they collectively posses the mandate to progress beyond mere reactions to the initiation of actions to positively manage change. Chapter 2 of this report aims to illuminate global and European drivers in relation to quality of life and the roles of cities and towns, regions, states and the EU, and considers:

- demographic development;
- changing consumption patterns;
- urbanisation.

This section further highlights related environmental challenges critical to quality of life and urban areas, including:

- air pollution;
- noise;
- climate change.

Chapter 2 also demonstrates the interlinkages with European policy as exemplified by European cohesion policy.

This selection of drivers, challenges and policies is clearly far from complete. Cities face many challenges. The selected examples aim to explore the effects of the processes listed above on quality of life, and in particular, on the quality of place, with a healthy environment as one of the prime requirements. The key drivers of change demonstrate the extremely complex and multiple interlinkages between all levels of governance in Europe.

Each section provides a description of the driver or challenge related to the urban situation and its interlinkage with European, national and regional levels, identifies gaps and barriers for more efficient policy-making and describes options for action.

2.1 Demographic changes

In 2008 the *International Herald Tribune* included the following quote from a speech by Miklos Soltesz, member of the Hungarian Parliament, on 9 September 2008 'The demographic situation in Hungary borders on the catastrophic, threatening our economic sustainability'. According to Eurostat, Hungary's population is expected to decrease by 13 % by 2060. East German cities and towns have already experienced similar or even greater population losses, changing life in these areas dramatically.

Trends like these, in tandem with general ageing of the population, decreasing household size and migration, are similar across many parts of Europe, influencing the material conditions and quality of life in cities and towns as well as people's needs and expectations. However, this section demonstrates that at each policy level, from local to European, it is possible to influence these demographic trends or their impacts and take action quickly.

Nature of changes

By 2065 almost one third of the EU's population will be older than 65, according to a forecast published by Eurostat (2008a). The combination of trends in fertility, life expectancy and migration will leave the total population size roughly unchanged by 2050, but will transform Europe's population structure. The number of young people in the EU will continue to decline; the population of working age will peak in 2010 but subsequently decline until 2050. Within this overall European picture of general trends there is, of course, significant variation at the regional level (see also Berlin-Institut, 2008).

The proportion of Europeans living in urban areas is set to increase from the current Figure of around 75 % to around 80 % in 2020 (EEA, 2006a; UN, 2008). In the short term, most of the increase will be due to rural to urban migration, but increasingly urban areas will experience immigration also triggered by the effects of climate change (EC, 2008a). However, cities and towns all over Europe again demonstrate local variations within this overall pattern. According to the *State of European cities* report (EC, 2007b), a third of cities grew between 1996 and 2001, a third witnessed stable populations, and a third experienced a notable decline in population (Map 2.1). In general, large cities have been expanding more quickly than smaller ones. Growth has been greatest in peripheral urban areas, while core cities within these urban agglomerations have experienced a decrease in population.

Population changes at national and local levels correlate mostly; however, statistics show variations between cities (Figure 2.1). These results indicate that there is opportunity for local policy to influence urban population development, at least partially.

Urban population mix

Also, at the micro scale of single cities, the composition of population groups has changed and will continue to do so. The *State of the European cities report* (EC, 2007b), based on audits of more than 250 cities, shows that the number of elderly people (65+) rose overall in most European cities with few exceptions. Cities with the fastest population growth are those with the smallest share of elderly people. However, in many Mediterranean cities population growth has continued in parallel with an aging population due to retired newcomers: the 'sun seekers'. Many central and eastern European cities have comparatively few elderly residents and many children. This may be due to the high birth rates of the late 1980s, but it is expected that in the future these cities will also follow the general European trends.

Figure 2.1 Population change in cities compared with national change between 1991–2004

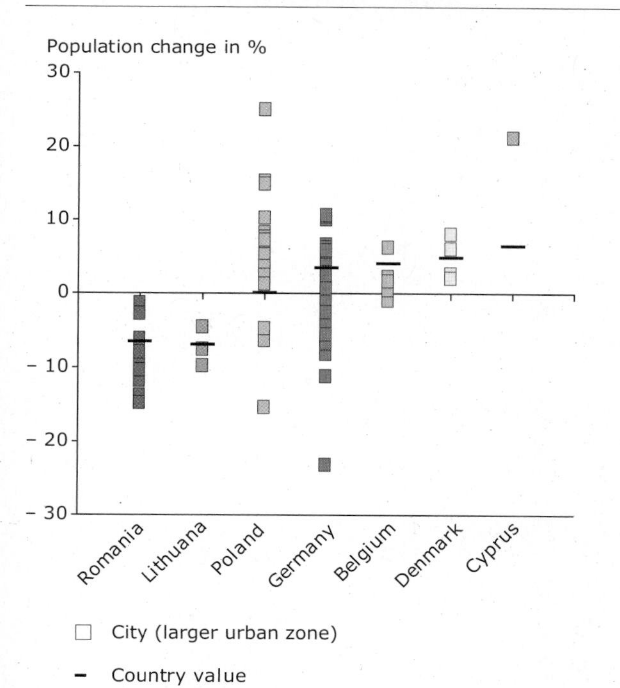

Source: Eurostat, Urban Audit Database.

Throughout Europe there is also a trend towards smaller households, and therefore more households. Household size is smallest in northern Europe (1.6 in Stockholm), slightly larger in Central and Eastern Europe and highest in Southern Europe (up to 3.4). Developments in cities show that one-person households gravitate towards urban centres, while in most cities families with children are leaving the urban core and settling in the surrounding suburbs.

Migration and immigration affects all cities across Europe (Map 2.2). In general, larger cities have higher immigration rates than smaller cities, which attract newcomers mainly from surrounding areas. Around three quarters of migration takes place within national borders. However, the percentage of non-nationals is rising, in particular in bigger cities, especially in Spain, Greece and Northern Italy. In part this is attributed to wealthy retired migrants from north-western European countries, who, attracted by nature, culture and mild climate, settle there on a more or less permanent basis, but it is also due to work-oriented migrants from poor countries in and outside the EU seeking work in the tourist industry along the Mediterranean coast (ESPON, 005b). Migration and mobility are likely to have an even greater role in urban population change in the coming decades.

Drivers of demographic change

Demographic development in cities is driven by many factors that affect economic, cultural, social and environmental dimensions of quality of life in Europe and its regions. Globalisation, rising mobility and continuing high population growth in Europe's immediate neighbourhood, especially in Africa, combined with poor economic performance and political instability may fuel further immigration (EC, 2007c). The impacts of climate change may also further reinforce these developments. At the European level, cohesion policy, economic policy, social policy, immigration policy and responses to globalisation influence these demographic trends. Similarly, at local level, quality of life is determined by policies relating to socio-economic development,

Map 2.1 Urban growth and population development 1990–2000

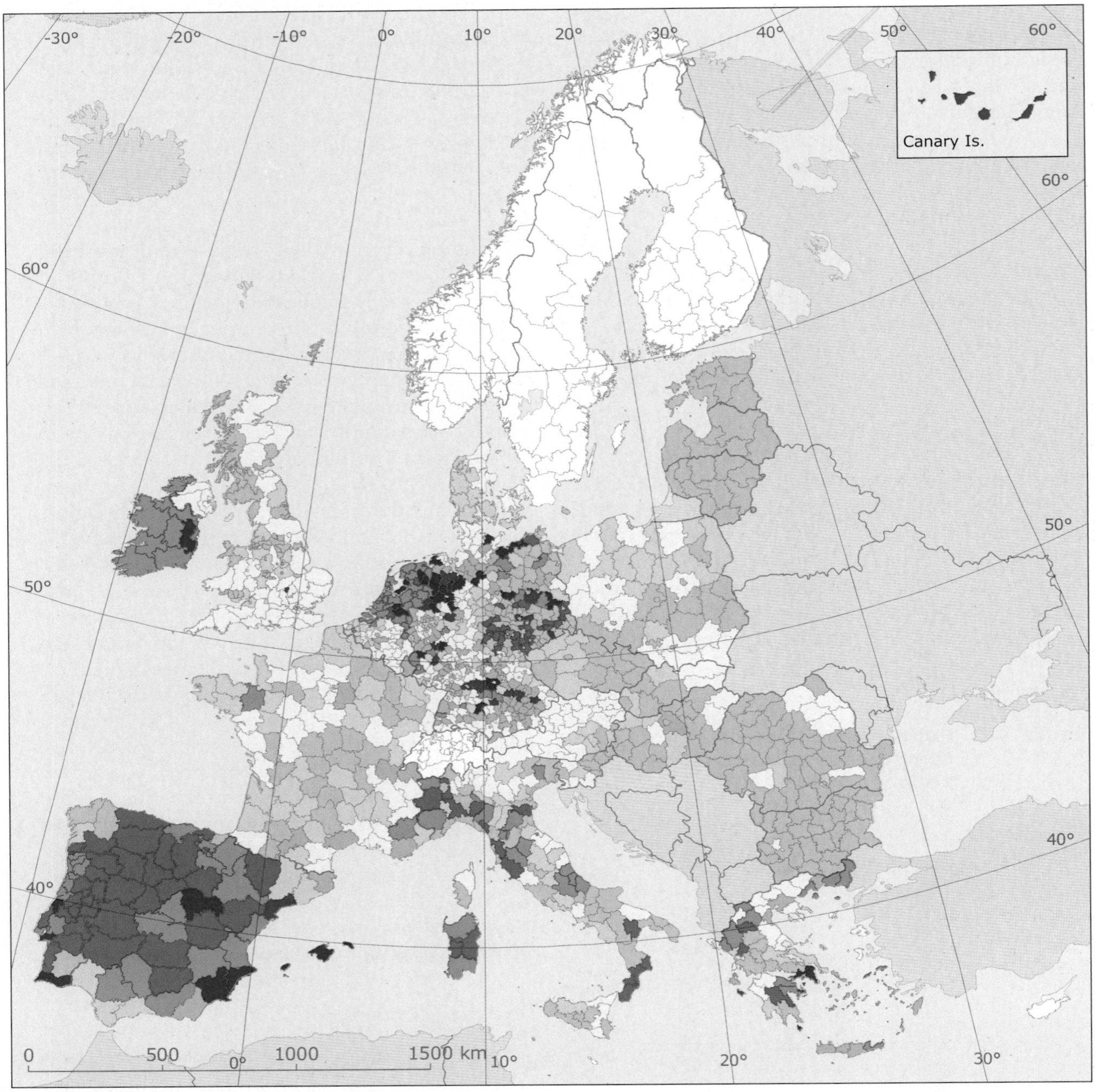

Urban growth and population development 1990–2000

Decrease of population density	Increase of population density		
Population + < Urban +	Population + > Urban +	+	Increase to below 10 %
Population + < Urban ++	Population ++ > Urban +	++	Increase 10 % and more
Population ++ < Urban ++	Population ++ > Urban ++	-	Decrease to below 10 %
Population - < Urban +	No data	--	Decrease 10 % and more
Population -- < Urban ++	Outside data coverage		

Source: GeoVille GmbH. Produced in the frame of ESPON 2.4.1.

Map 2.2 Urban Audit cities — number and origin of newcomers, 2004

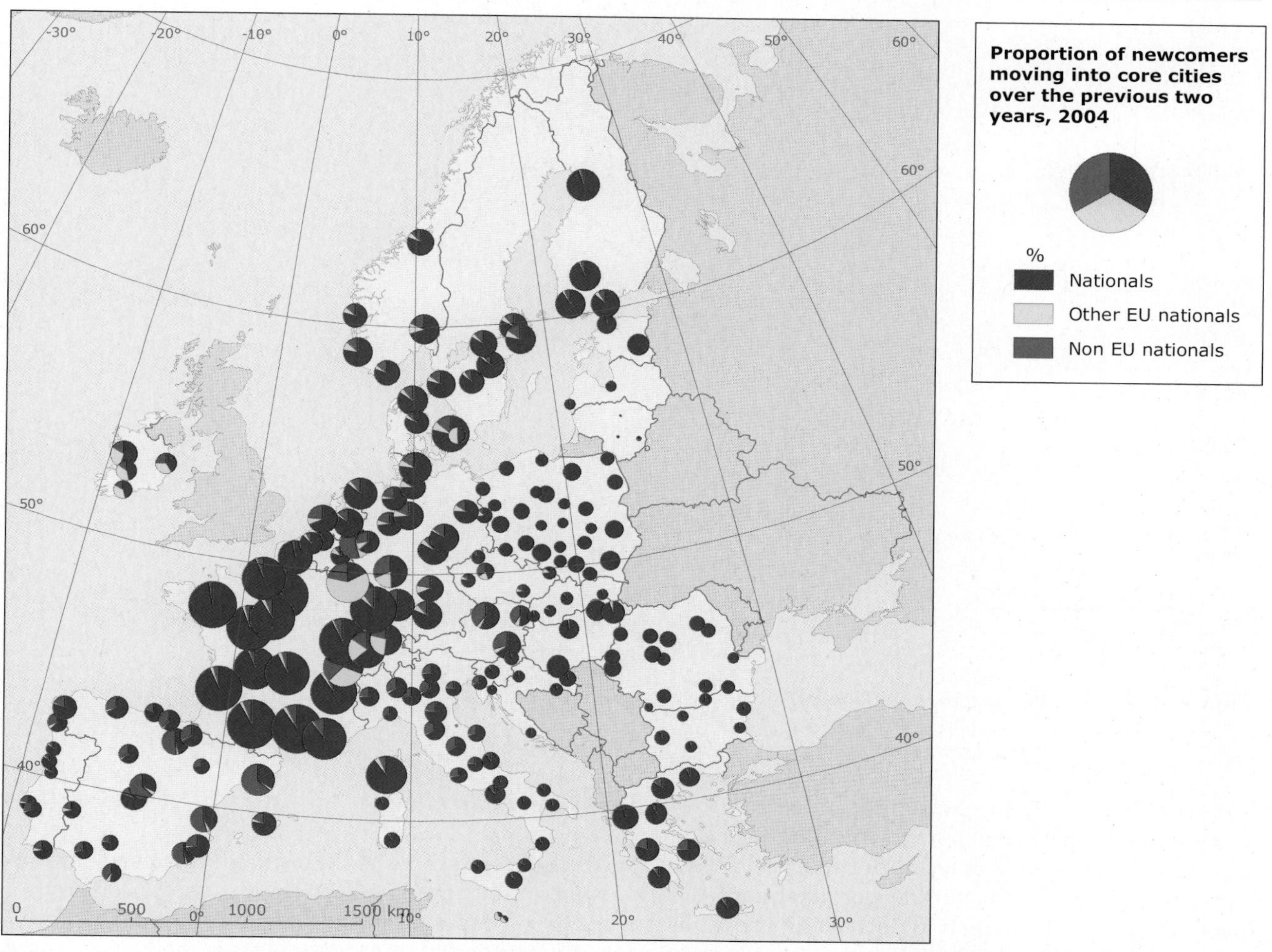

Note: For Austria, Bulgaria, Czech Republic, Croatia, France, Hungary, Italy, Latvia, Poland and Portugal the data are from 2001.

Source: Eurostat, Urban Audit Database.

culture, urban planning and design, and affordable housing, and also by the environment, for example good air quality, low noise levels and access to green space.

Quality of life determines whether population groups — the young, elderly, families, immigrants, poor, rich etc. — are attracted by the city and decide to live there; or, if conditions are unfavourable, they leave. Cities therefore have, through their policies, the potential to influence European and global demographic trends in their locality.

Impacts of growth and decline

When the urban population grows, land-take increases as does consumption of energy, water, material and food. All this is potentially harmful to the environment, and may contribute to or inhibit sustainable development and quality of life (see Section 2.2). At the same time, a higher urbanisation level and relatively high population densities offer the possibility of living more efficiently with respect to energy, water and urban land use per inhabitant. Cities are also transport-energy efficient, as demonstrated by London's low transport footprint compared to that of other English cities (see Section 1.4) and as shown in the graph of energy consumption in cities (Figure 2.2). People living in densely populated areas are more likely to walk, cycle and use public transport (UITP, 2006). In short, population growth in cities will increase cities' impact on the environment, but a higher proportion of people living in relatively dense urban areas offers potential for increased sustainability.

In contrast, shrinking cities face different problems. Economic and social activity is normally decreasing and there is generally lowering of consumption and its related pressures on the environment.

Figure 2.2 Energy consumption for passenger transport versus density

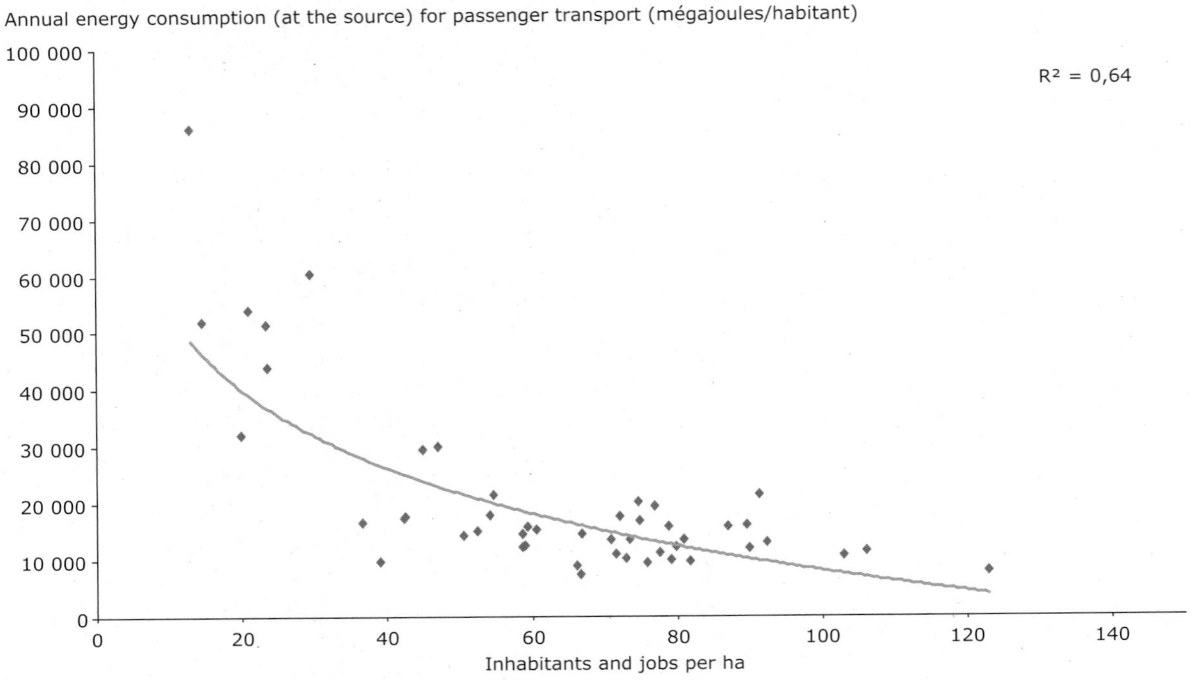

Annual energy consumption (at the source) for passenger transport (mégajoules/habitant)

R^2 = 0,64

Inhabitants and jobs per ha

Source: Mobility in Cities Database, © UITP, 2006.

However, whilst there is no significant reduction in urban land use, urban sprawl continues, leading to a reduced efficiency in terms of urban land use, transport, energy and water use per inhabitant. Declining populations lead to decreasing use of the existing infrastructure, which then has overcapacity. Underuse of water and sewage systems and the extended storage of water and sewage in the pipe systems can causes hygiene problems. Additional flushing necessary to maintain the infrastructure results in higher water and energy use per capita. As another example, vacant flats in apartment buildings lead to higher energy consumption — up to 31 % more than for a fully occupied building (UBA, 2007). Shrinking cities and towns can no longer effectively provide services like schools, hospitals and shopping. Also, public transport becomes inefficient, resulting in more use of cars, which in means that social groups are excluded from services as well as employment.

Population changes and consumption

Different population groups — old and young, poor and rich, native and immigrant — have individual lifestyles, and different ideas, perceptions and expectations of quality of life. All influence urbanisation and consumption patterns (see also Sections 2.2 and 2.3).

Smaller households

The number of smaller household European cities is increasing. Smaller households tend to consume more resources per head than larger ones, as demonstrated by the example from the Netherlands (Figure 2.3). However, in regions of declining population, the number of households may remain unchanged. If specific measures to restore areas to open space are not put into place, there is no consequent reduction in urban land and living space.

Households with fewer members also tend to use more energy and water per person. For example, in the United Kingdom the water consumption per capita is 40 % greater in single households compared to that in two-person households and 73 % greater than in four-person households (POST, 2000). Also, although the number of persons per household is declining, the average living space in new dwellings is tending to increase across Europe (Figure 2.4).

Older people

The consumer behaviour of older people is not the same as that of other younger groups as they generally have different needs and different financial and physical capacities. For example, some

Figure 2.3 The Netherlands — expenditure per head according to household size, 2000

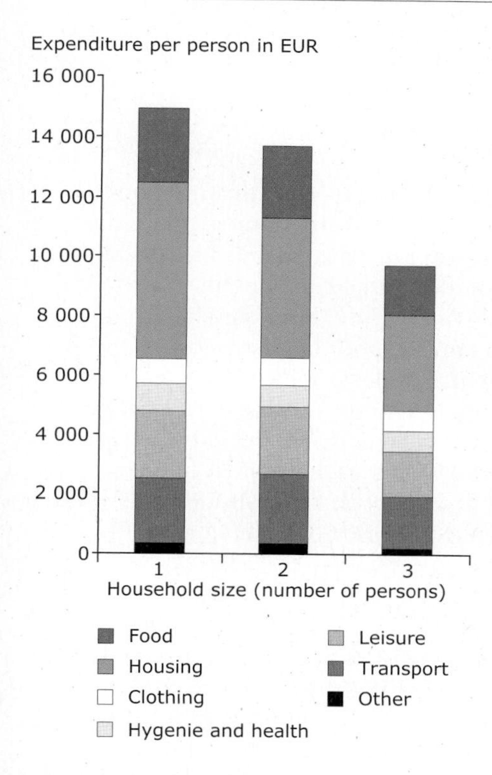

Source: CBS, 2004.

older people buy second homes and move over long periods to mountain or coastal areas, which are particularly vulnerable in terms of environmental pressures. According to a study in the United Kingdom (UK Office for National Statistics, 2007), the proportion of household expenditure on food, drink and transport rises with age. In France people over 65 spend 19 % of their income on these items compared to the national average of 15 %.

However, it is anticipated that the next generation of elderly people will follow a different pattern of consumption. The baby boomers of the 1980s will probably continue to drive cars when retired. Eurostat statistics show, that older people on holiday make on average the most and longest trips (Eurostat, 2008b). Nevertheless, a sizeable proportion of the older population will have some type of disability, and will benefit from alternative transport modes that meet their individual needs supported by intelligent land-use planning to provide the necessary services and facilities for elderly people in their neighbourhoods (OECD, 2001).

Immigration and migration

Movement of groups with other national or non-European backgrounds contributes to new consumption styles by introducing aspects of the migrants' own cultures to their new community. Social and cultural factors largely determine lifestyle and individual behaviour. Immigrants have different cooking habits and buy different foods; they may have different household sizes and demands on living space. They also spend their leisure time in different ways, for example they may be less likely to travel abroad but more likely to meet frequently with family and friends, and have special needs for public open space. Meat consumption in India, for example, is far below the levels of western Europe and pig meat is not eaten at all in Turkey (Table 2.1) — behaviours that immigrants bring to Europe. In southern Europe the average household size is larger than in northern Europe due to people's socio-cultural background. Although this is beginning to change, for example Portugal, Slovenia and Spain have recently experienced the largest decreases in household size (EC, 2007b).

Figure 2.4 Rising average size of newly completed dwellings

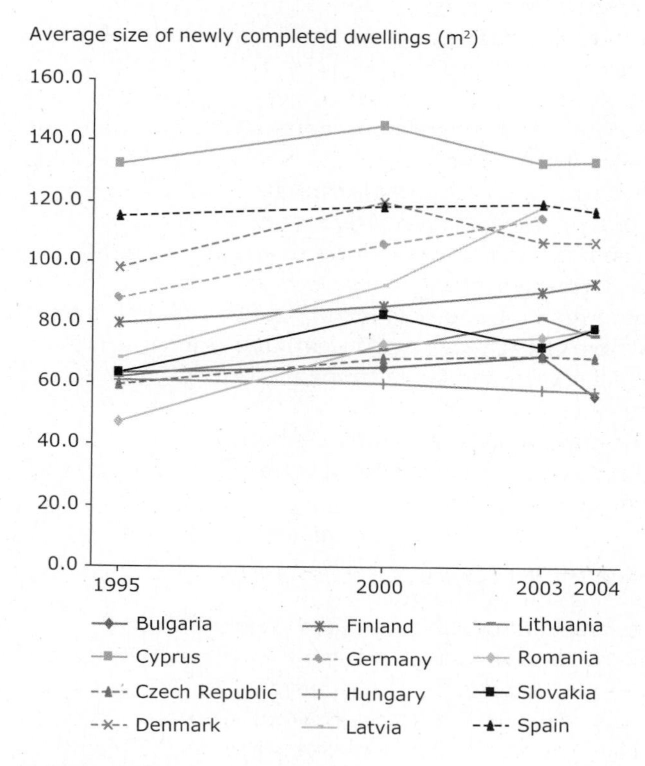

Source: UNECE, 2006.

Table 2.1 Consumption of selected foods in 2003

	Western Europe	Turkey	India
Meat in kg/capita/2003	90.86	20.57	5.23
Pig meat in kg/capita/2003	43.95	0.00	0.46
Milk products in kg/capita/2003	256.50	122.25	67.99

Source: FAOSTAT, 2003. Statistical database of the UN Food and Agriculture Organization http://faostat.fao.org/.

Changing demographics

Many more groups based on gender, families versus single or childless couples, rich and poor households, highly educated etc. can be identified and show different consumption behaviours at the micro scale. Consumption patterns change as demography shifts, but also overlap and are further mixed by European and global trends as well as general trends in lifestyle and individual choice. As cities show different responses to the general demographic trends, they also consume in different ways, and have different impacts on the local environment and quality of life.

Policy for demographic change

Cities and towns in Europe can avoid the negative environmental impacts of demographic change, and reinforce the positive effects, by:

- influencing the demographic changes in a positive direction;
- adapting to inevitable changes.

The European Commission's Communication *The demographic future of Europe — from challenge to opportunity* (EC, 2006a) identified five key policy areas where constructive responses to the demographic challenges could be developed:

- promoting demographic renewal;
- promoting employment in terms of more jobs and longer working lives of higher quality;
- promoting a more productive and competitive Europe;
- receiving immigrants;
- sustaining public finances to guarantee adequate social security and equity between the generations.

The European Union therefore supports the Member States as part of a long-term Strategy to address these challenges. Recommendations are based on assessment of the impacts on the labour market, productivity and demographic growth, as well as on social security and public finances. The assessment and related policy are oriented towards the stabilisation of the socio-economic situation in Europe but do not consider the environmental aspects of demographic change.

Cities are also beginning to develop strategies to respond to current and future demographic changes. Different measures aim to attract certain population groups by providing financial benefits, for example affordable flats or tax reductions for skilled workers from abroad, and the provision of services or appropriate urban design, including green open space or playing grounds. Eastern Europe, in particular East Germany, having experienced massive population losses, has gained much experience in dealing with these situations and other cities can learn from this (Box 2.1).

Policy gaps

More and more consideration is being given to demographic change in European, national and local policy-making. As local trends can differ significantly from general regional or European trends, European policy to influence demographic change will sometimes not meet local needs and, rarely, may be counterproductive.

Environmental policy hardly ever deals with the negative effects of changing demography and consumption. Typically, it focuses on single environmental issues and is seldom connected to policies influencing demographic change. Discussions on the consumption patterns of different social groups and their environmental impacts have only just begun, mainly at the level of research.

Overcoming barriers to action

Demographic changes and their impacts are not only driven by European and global trends. Europe and its cities and towns can aim to influence demographic changes in positive directions and facilitate adaptation to inevitable changes.

At the same time, it is clear that not all demographic developments can be driven in the desired directions. The situation is complex as it is not always possible to determine whether specific demographic developments are favourable or not. However, from a local perspective, environmentalists need to link their work with other policy areas to actively drive demographic

Box 2.1 East German Bund–Länder-Programme — urban conversion

Initial situation

After German reunification in 1990, many regions in East Germany experienced a massive population loss. This overlaps with the general aging process and population loss in Germany as a whole. The supply of flats in such areas is much higher than the demand and has resulted in high vacancies especially in big apartment building blocks. This situation has the potential to lead to deprived neighbourhoods.

Solution

The German government and the German Länder have initiated the Stadtumbau Ost (Urban Conversion East) programme, which aims to improve the attractiveness of the East-German cities and towns. It aims to support the renewal of the town centres, the reduction of the oversupply of flats, and the revaluation of cities which are affected by shrinking processes. The demolition of 350 000 of the 1 million flats that remained unoccupied until 2009 is planned in municipalities where there is an above average vacancy rate and which have a revitalization concept. At the same time, the municipalities will take actions for renewal, such as the revaluation of existing buildings and quarters with particular cultural and historical value, the adaptation of the urban infrastructure, the reuse of areas of open land, and the improvement of neighbourhood quality.

An amount of EUR 2.5 billion was available for the years 2002–2009. So far 342 municipalities took part in the programme between 2002 and 2005.

Results

Example Dresden Gorbitz:

Development of a quarter with large apartment blocks to a quarter with more differentiated and attractive structures.

Before

Photo: © Archive photos of Eisenbahner Wohnungsbaugenossenschaft Dresden e.G.

After

Photo: © Archive photos of Eisenbahner Wohnungsbaugenossenschaft Dresden e.G.

Example: Aschersleben — contraction of the settlement

Ascherleben's strategy is to downsize the town after a decrease in population of 19 % between 1990 and 2000. The town renovated central areas and increased their attractiveness while demolishing large areas of unattractive peripheral multi-storey apartment areas dating from East-German times. Although demolition is costly it will save money in the long run by the removal of underused infrastructure and the improvement of the town's general appearance and image. Furthermore, the town relocated schools and other administrative offices back into the centre.

Photo: © Stadt Aschersleben

Source: http://www.stadtumbau-ost.info/. http://www.aschersleben.de/index.asp?MenuID=27.

development in the desired direction to ensure quality of life not only in economic terms. In all other cases, European and national governments, as well as the cities, must monitor demographic developments. These procedures will permit timely adaptation to changing needs and the opportunity to develop accompanying measures, which aim to reduce environmental pressures and ensure quality of life.

Cities and towns

Urban areas can provide favourable environments and enhanced quality of life for certain population groups, for example by providing high quality child care and creating safe and child friendly environments for young families. In general, the greater the diversity of social groups within urban society, the greater the potential for the realisation of the long-tem sustainable development of the city. To attain these objectives, municipalities need to analyse local demographic development and the needs of different population groups, so that they can determine the best strategies for sustainable development.

Shrinking cities can create greener and safer single-housing areas to encourage population retention. By preventing urban sprawl, such cities can create urban quality and compactness, and so become more transport and energy efficient (see example in Box 2.1). At the same time it is necessary to adapt the technical infrastructure and services to population decline.

Growing cities and metropolitan areas will similarly need to investigate the needs and interests of their new population groups, and develop strategies to manage growth in a sustainable way, based on an integrated approach to local sustainable development (see also Chapter 3).

Policies on a wider scale

European and national policies to influence demographic change require more differentiation and effective connection with regional and local policies.

European policy needs to integrate the environmental impacts of demographic change within its policy frameworks. These impacts differ widely at the regional and local levels, and so a much more differentiated policy is required to cope with the variety of demographic changes and their impacts on the environment. For example, the focus cannot only be on population growth or stabilisation; policy must also allow for population decline in some regions and manage these declines to secure the most positive outcomes. The inclusion of environmental impacts in demographic assessments at the European level, like the biannual assessment of Europe's demographic future (EC 2007c), would help with this.

More Europe-wide research is needed to identify the trends in consumption patterns of the various demographic groups across Europe, and to identify the potentials within each group to reduce environmental pressures. Guidance should be provided to cities and towns to develop sustainable consumption patterns by considering and adapting to demographic changes.

2.2 Consumption and urban lifestyles

Consuming food, buying clothes, having a warm and dry shelter are indispensable for our lives. A higher income enables us to buy more food and clothes, bigger apartments and many other goods and services; meaning not one only television for the family but others for the children, the bedroom or the kitchen. These new goods and services can provide us with more choices and a means to enjoy our lives more fully. In the search for a higher quality of life we push the limits further and further. We can travel further to enjoy remote and new places, yet even the Antarctic is no longer remote. In doing so we consume more, highly processed food, and travel everywhere by car, resulting in alarming rates of obesity and serious health problems. So is consumption really providing us with a new and better quality of life?

This section demonstrates the importance of consumption in urban lifestyles as a socio-economic driver that significantly influences the possibilities for a more sustainable quality of life in cities, and the ways in which inappropriate consumption can undermine quality of life. Action at the individual level to secure more sustainable forms of consumption is critical in providing decisive contributions to collective efforts. If governments offer citizens the opportunity to live sustainably, the possibilities for the realisation of improved quality of life are enhanced.

Consumption provides quality of life

Consumption is the use of goods and services to fulfil our basic needs and demands. As such it is crucial for quality of life. People consume energy, resources, food, water, and land for nutrition, housing, mobility, recreation, communication, education and entertainment. In addition to private consumption, public and business consumption have an equally important role. A high level of material consumption is generally seen as an indicator of advanced development and well-being, therefore the conventional economic development model builds upon rising consumption.

European consumption is rising as measured in terms of the expenditure of households (Figure 2.5) and public entities on goods and services.

Figure 2.5 Changing household consumption patterns in EU-10 and EU-15

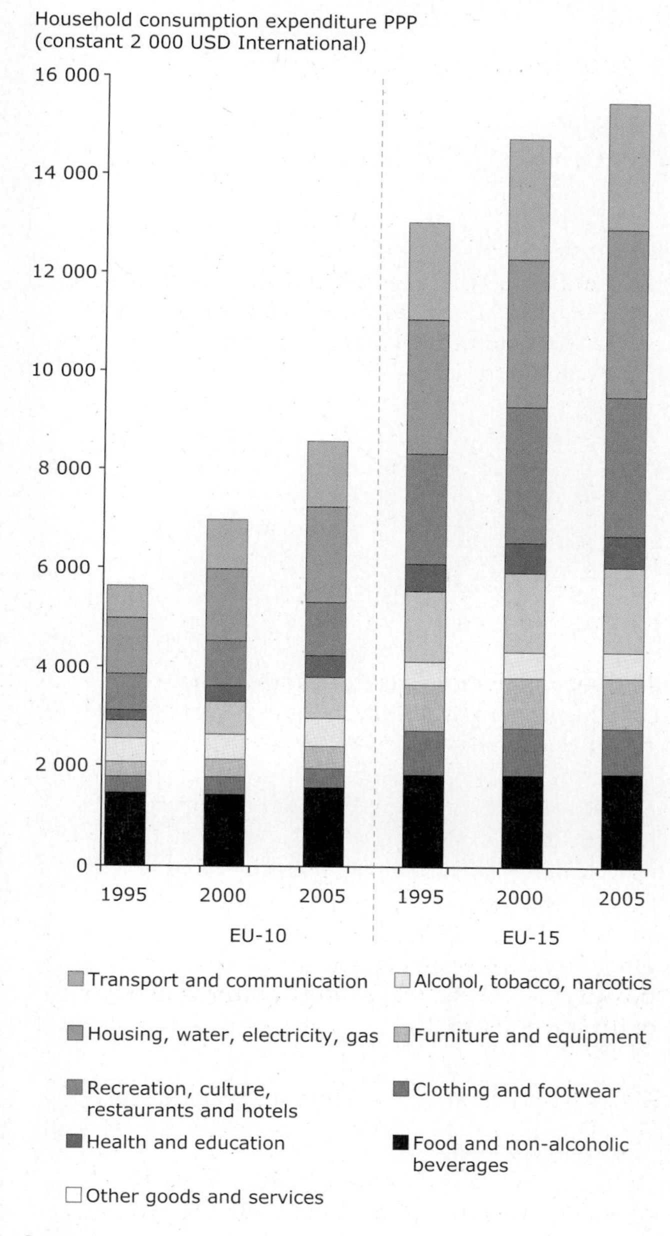

Source: EEA, 2007b.

Many factors drive consumption patterns in Europe. A major factor is the availability of income and budgets that determine to a large extent the availability and affordability of goods, infrastructures, services and technologies. Most people tend to follow general societal trends. Moreover, business, in order to maximise profit, creates new markets and stimulates demand beyond basic needs.

Trends across Europe

Lifestyles and related consumption are also influenced by culture and history as well as individual values and preferences. The growth of the wealthy middle class throughout Europe contributes to changing values and the associated consumption patterns, also called 'hyper-consumption' (Lipovetsky, 2006) and 'age of access' (Rifkin, 2000). Consumption has shifted from the simple purchase of particular products to the purchase of integrated packages that provide access to specific experience. For example, people charge products such as cars with symbolic and social meanings, primarily related to social status, making the statement — 'I am rich'. Today's consumers are also paying for access to the meaning beyond the product and thus buy an entire lifestyle and identity associated with a particular product. They choose houses, holidays and cars to make a statement — 'I am dynamic', 'I am smart' etc. These cultural trends are reinforced by business strategies, and often result in increasing material consumption.

Policy influences trends

Beyond these general trends, and because of different social patterns, cultural and governance traditions as well as the economic situation, the way we consume and the related impacts differ substantially across Europe, its regions, cities and towns (Box 2.2). So, for example, even cities of similar size and wealth show substantial differences in transport choices — from 58 % of the population using public transport in Prague to fewer than 25 % in Naples and in Rome, or from 29 % using the bicycle in Copenhagen to fewer than 5 % in most other cities (Ambiente Italia, 2007). Furthermore, demographic characteristics influence consumption both now and in the future (see Section 2.1).

European, national and local policies can actively influence these drivers, for example the availability of products and services or prices, or increasing consumers' awareness of the (hidden) environmental and social costs of products. However, other policies can also have indirect and unintended effects on

Box 2.2 Differential consumption patterns in urban Europe

Map 2.3 Average floor space per resident in core cities, 2004

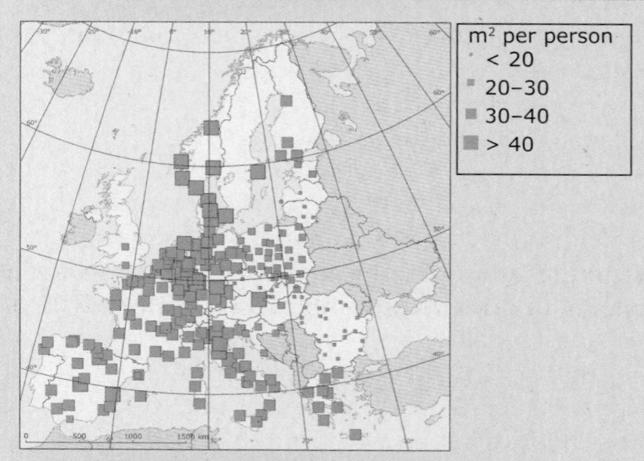

Note: For Belgium, Bulgaria, Czech Republic, Croatia, Estonia, France, Hungary, Italy, Latvia, Norway and Switzerland the data are from 2001.

Source: Eurostat, Urban Audit Database.

Figure 2.6 Car registration rates and travel to work by car

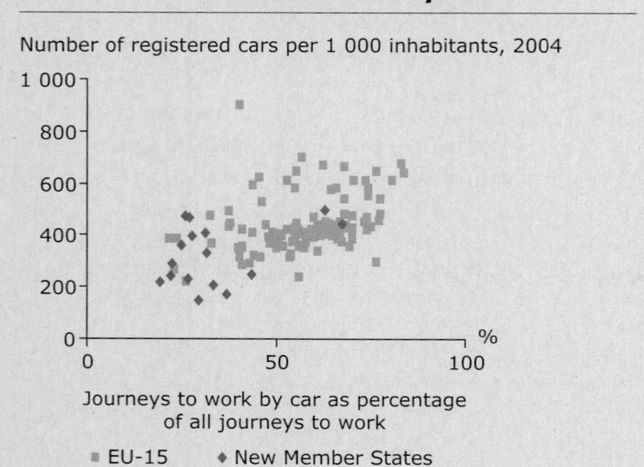

Note: For Austria, Czech Republic, France, Italy, Slovenia and Switzerland, the data are from 2001.

Source: Eurostat, Urban Audit Database.

Figure 2.7 Public transport passengers

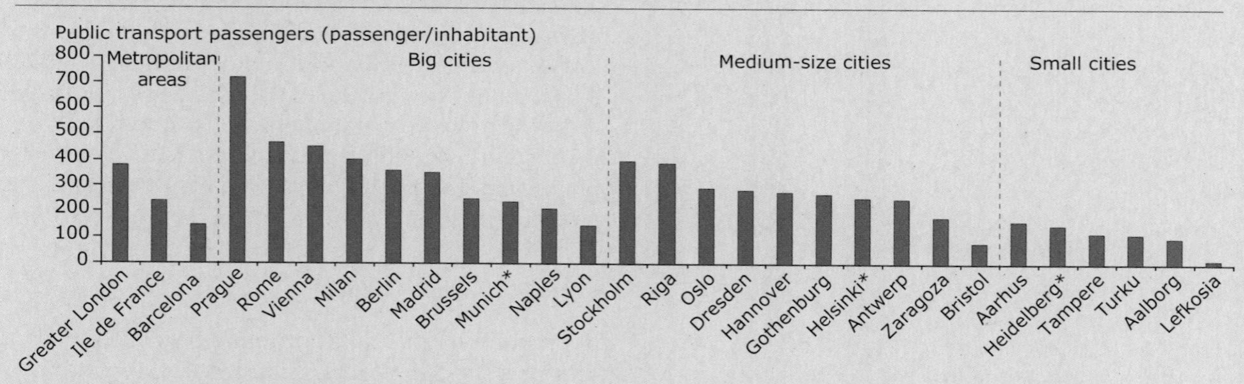

Note: * Figure refers to a public transport network that serves twice the amount of the city's population.

Source: Ambiente Italia, 2007.

Figure 2.8 Per capita urban waste production

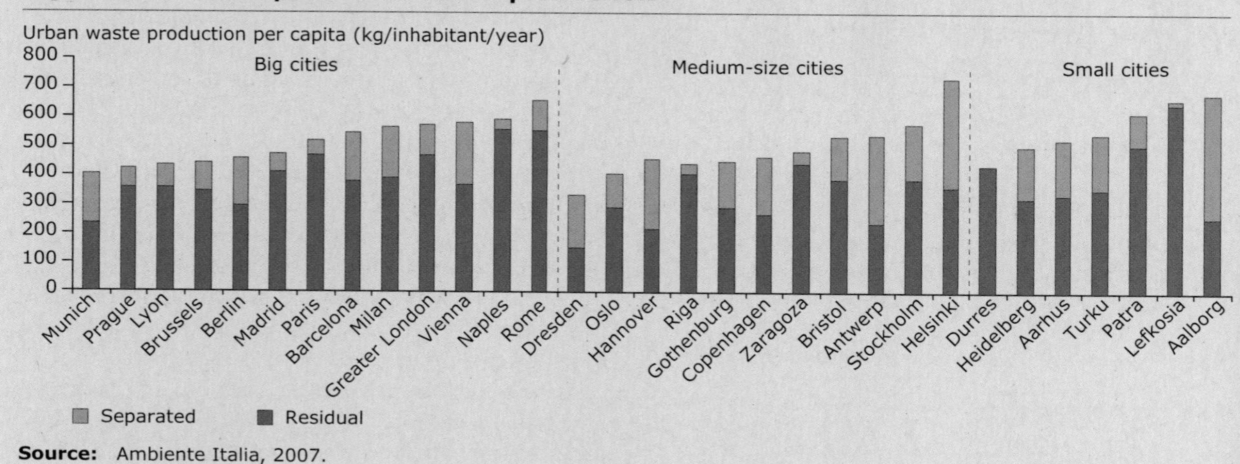

Source: Ambiente Italia, 2007.

consumption patterns. For example, European cohesion policy will contribute to higher incomes in the new member states in the longer run, and so alter consumption patterns there. Similarly, if local communities focus only on new or upgraded roads while neglecting public transport needs, people are indirectly encouraged to use the car.

Paradox of affluence?

There is no question that the production of goods and services and their consumption provide essential elements of quality of life. The problem is the fact that consumption can equally have negative impacts on other facets of quality of life, for example:

- social impact, for example by excluding low income households from certain goods or services;
- environmental impact due high usage use of resources such as land, energy, water, materials (including food) and the generation of waste and emissions such as air pollutants, noise and greenhouse gases.

As a consequence, there will be health problems, economic losses and social inequity. As an example, energy use contributes in several ways to a higher quality of life in cities. It illuminates and heats our homes, shops, public buildings and streets and enables the supply of public services. On the other hand, energy generation based on fossil fuels causes considerable environmental pressures in the form of emissions of greenhouse gases and acidifying substances.

Of the consumption categories illustrated in Figure 2.5, around two thirds of environmental pressures such as greenhouse gas emission, material use, ground-level ozone-forming emissions and acidifying substance emissions arise from the consumption of food and drink, housing and mobility (EEA, 2009a). Although Europeans consume differently in the various regions, the overall consumption patterns and the associated use of resources and generation of emissions are not sustainable (Box 2.3). Furthermore, consumers in other parts of the world are increasingly adopting European lifestyles. Figure 2.5 shows that levels of consumption are not only generally rising in the EU, but also that the consumption patterns in the new member countries are approaching those in old Member States, reflecting a change in lifestyle and a general increase in disposable income (EEA, 2007b).

Impacts on urban environment

Apart from overall consumption patterns, the way city dwellers prefer to live, enabled by the organisation and design of their city, influences the urban environment in many ways; for example, different urban transport systems and methods in place for the delivery of goods can have different impacts with respect to urban air quality and noise emissions (see also Section 2.4 Air pollution and noise, Box 2.13). Transport-related problems will be generally greater in cities with a high proportion of individual motorised transport, compared to cities with good public transport and high levels of walking and cycling. The scale of the problems also depends, of course, on the car fleet and city design. Cities with well-organised systems for separate collection of waste have lower impacts on the environment than those relying mainly on landfill. Compact cities, where most people live in multi-storey buildings, take up less land per inhabitant than cities where

Box 2.3 European consumption patterns

- Around 15 tonnes/capita of materials (fossil fuels, biomass, metal ores, minerals) are used each year to produce the goods and services (including energy) that we consume in Europe, and this amount is expected to grow by around one quarter by the year 2020.
- At the same time, the amount of municipal waste in the EU is also expected to grow by around one quarter. Even with better recycling and less landfilling, the overall growth in waste amounts still poses a major challenge.
- The constant growth of consumption volume often outweighs environmental efficiency gains, e.g. the total fuel consumption by private cars in EU-15 has grown by 20 % between 1990 and 2004 in spite of fuel efficiency improvements of more than 10 % per car, due to the increase of kilometres travelled.

Source: Eurostat/IFF, 2007; EEA, 2007b.

single houses prevail. This conserves land for agriculture, forestry, nature and biodiversity, and housing and transport are more energy efficient (see also Section 2.3).

Global dimensions of consumption

Europe's citizens use many resources from locations far away from the city, and produce waste and emit pollutants and greenhouse gases that have impacts far outside the municipality, often in other parts of the world. For instance, urban traffic is responsible for 40 % of greenhouse gas emissions and 70 % of pollutants of European road transport (EC, 2007d). It is obvious that the geographical area of a city cannot deliver the necessary resources and services. However, depending on the local level of consumption, the ecological footprint [4] varies widely between cities and countries all over the world, in particular between the developed and the developing world. In 2005 7.5 % of the world's population lived in the EU-27 — mostly in cities — but generated 13 % of the world's ecological footprint (WWF, 2008).

Consumption in European cities is high; 69 % of all energy consumed is used in cities. However, the average urban dweller consumes only 3.5 million tonnes of oil equivalent (Mtoe) in relation to the 4.9 Mtoe consumed by a rural dweller (IEA, 2008). Exploiting this potential is key to more sustainable development. Finally, the extent to which consumption contributes to or threatens quality of life depends on choices taken and the level of consumption. More sustainable consumption ensures quality of life now and for future generations and reduces social inequalities.

Europe sets the frame

In July 2008, the European Commission published an action plan on sustainable consumption and production, and sustainable industrial policies. This action plan includes proposals to make products and services more sustainable, for example by extending the Directive on the eco-design of energy-using products to more product categories, reviewing the Eco-labelling Directive and the Energy Labelling Directive, as well as establishing a harmonised basis for green public procurement.

Other EU policies seek to influence consumption; for example, the Directive on the energy performance of buildings, the Directive on Energy Efficiency and Energy Services, and the European Strategy for the prevention and recycling of waste, all of which can have considerable influence at the local level.

Urban policy makes a difference

Some cities already actively aim to secure more sustainable consumption patterns based on the application of a variety of policy instruments, including Local Agenda 21 processes initiated in more than 5 000 European municipalities (ICLEI, 2002).

Some cities implement 'demand-side management schemes', with actions focused on providing choices for more sustainable consumption, and stimulating behavioural changes. Such measures include more pedestrian areas, bicycle lanes, car- and bike-sharing schemes, public transport integrated fares, parking and congestion charges. Other cities actively support the uptake of renewable energies and promote energy efficiency, directly involving individual consumers (Box 2.4). Separate waste collection and green public procurement schemes also aim to make citizens as well as local government officials aware of sustainable products and lifestyles. However, all these existing solutions require widespread implementation. To date they have only been applied by a few pioneering local authorities.

Barriers to effective policy-making

In general these approaches are only the first step towards the reduction of the impact of Europe's high and unsustainable consumption. Our current model of economy and social welfare still builds to a large extent on rising consumption and GDP growth. Success in realising more sustainable consumption will be limited as long as this paradigm remains unchanged. An alternative model of smart growth can ensure a socially balanced quality of life in the longer term, based on the integration of policy fields including those relating to economy, social and demographic issues.

Distorted prices and limited or ineffectual choices produce the wrong incentives. For example, in the case of land take for housing, many stakeholders — local government, land owners, regional planning authorities, land developers, banks, households, and infrastructure providers — all take their own individual and too often unconnected decisions (see also example in Box 2.11). All these single decisions make full economic sense from the individual

(4) 'The ecological footprint measures humanity's demand on the biosphere in terms of the area of biologically productive land and sea required to provide the resources we use and to absorb the waste.' (WWF, 2008).

Box 2.4 Nyíregyháza (Hungary) — improving housing energy efficiency

Initial situation
Nyíregyháza is situated in the north east of Hungary. It is the 7th largest town in the country with 120 000 inhabitants. Almost one-third of the cities' housing stock was built using concrete panels in the 1960s and 1970. The energy consumption within these buildings is extremely high: they suffer from very poor insulation, with numerous thermal bridges, poor air-tightness and severe water infiltration. The depreciation of this building stock can also lead to severe social problems and to the creation of urban slums.

Solution
There are 44 000 households in Nyíregyháza, half of them in blocks of flats, of which 12 644 are connected to a district heating system. The city decided to modernize its district heating system and housing stock. In 1997 the first programme 'Opening', started to upgrade the thermal distribution circuits for more than 12 800 flats. In 2001 the 'Panel programme' involved retrofitting panel blocks. Most of the flats involved in the programme were privately owned, which presented a challenge in securing agreement to retrofit.

Results
The Panel Project has resulted in energy savings of 26.8 TJ/year. An evaluation of the retrofitting measures has shown that an overall energy saving of 68 % can be achieved. The projects were the most cost-effective measure possible.

Source: http://www.display-campaign.org/rubrique682.html.

perspective, but without coordination the result is typically urban sprawl with low population densities creating massive follow-up costs for society — an outcome nobody would have chosen.

Fragmentation of responsibilities as well as extreme decentralisation expressed in inappropriate institutional structures provide a further barrier. For example, energy-efficiency actions often occupy only a marginal place in Member States' Operational Programmes for the EU Structural Funds. One reason is that energy authorities are often responsible for energy supply and energy efficiency at the same time. This can lead to business conflicts and typically the authorities focus on energy supply rather than managing demand. Energy efficiency is often perceived as being more complicated to implement than measures for energy supply; thus hindering or delaying such measures.

People's perception can add further challenges. For example, by living in low density suburban areas, where housing and land is cheaper, individual households can save a lot of money. On the other hand, they need to invest more in transport to reach work places, schools or other services. These additional personal costs, apart from costs for reduced biodiversity and ecosystems services, typically fully neutralise the savings over a longer period; yet, this fact is often not perceived by households (Figure 2.9). In addition, the maintenance of infrastructure used below its capacity leads to higher costs per user. These costs must be met directly by all citizens in the form of higher charges or else the municipality has to pay. As a result, all society has to pay for the greater environmental costs. Overall, this leads to an unsustainable situation, economically, socially and environmentally (UBA, 2009).

Lifestyles are a domain of individual action highly influenced by policies measures. Thus, in spite of general public awareness of major environmental problems like climate change, only a minority of Europeans take individual action for ecological reasons (Eurobarometer, 2008). People expect change from the 'others' before changing themselves. This situation can persist if there is no action at the collective level. Often, policy intervention contributes to unblock the situation, for example the use of bicycles after the implementation of cycling lanes and the city bike system VELIB in Paris, or waste sorting in France adopted by more than 70 % of households after municipal equipment became available.

Overcoming barriers to action

Consumption patterns will continue to have adverse impacts on the environment and social equity, unless the society moves to lifestyles that use fewer resources and build upon ecosystem principles that emphasise, for example, the greater share of renewable resources.

It is the responsibility of policy at all levels to set the framework that provides the basic conditions for sustainable consumption. Only then will individual citizens be able to choose more sustainable lifestyles and meet their personal responsibilities. Equally, if cities and towns can provide high quality environments, which fulfil the needs of citizens for safe areas, green and other public spaces, as well as for short distances to facilities and services, then city centres can become sufficiently attractive to counter urban sprawl. In this way, cities can also reduce their energy and transport use and help protect areas outside the city for agriculture, recreation or wildlife without degrading the quality of life. The role of governments is to provide the right framework conditions, incentives, and facilitate the debate amongst different interest groups at the relevant levels.

Integrated multilevel policy-making

Every governance level, from local to European, needs to take responsibility and cooperate in order to develop horizontally integrated and multilevel approaches supported by appropriate management structures and appropriate governance. The crucial role of the local level to transform the European situation is often underestimated. Local authorities are indeed those responsible for renovating and developing new districts, managing land-use, planning and organising mobility, and thereby substantially influencing the ecological footprint of cities. Box 2.5 on energy-efficiency policies shows an example of how government action at multiple levels can be integrated in order to achieve success — in this example, to achieve the objective of greenhouse gas reductions and thereby mitigate dangerous climate changes.

Fundamentally, there is a need for broad participation by all relevant stakeholders, including citizens and NGOs in order to manage the inevitable conflicts between the interests of individuals and groups, and to meet societal needs for quality of life for all, now and in the long term.

Foundations of choice

Policy, based on coordinated initiatives at every level, needs to provide the right infrastructures to enable consumers to choose more sustainable lifestyles expressed in the consumption of sustainable goods and services, for example use of energy-efficient appliances, consumption of sustainably produced food, use of public transport systems, recycling of waste, etc. (see examples of urban action in Box 2.6).

Figure 2.9 Greater Hamburg (Germany) — modelled costs for transport and housing in residential areas

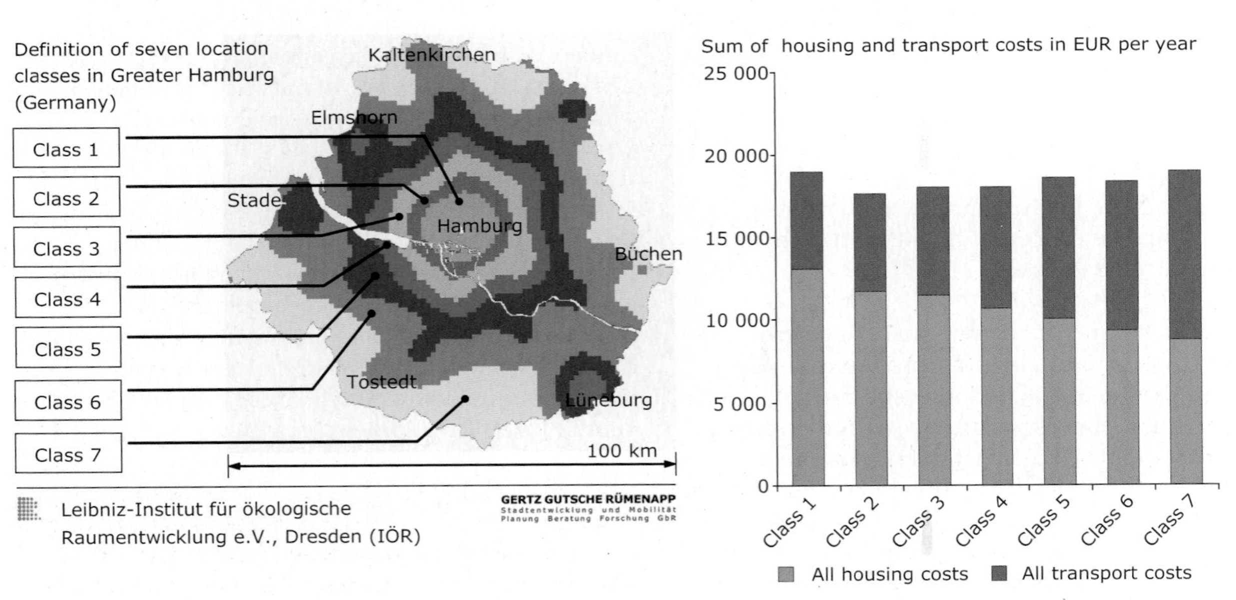

Source: UBA, 2009.

Box 2.5 Interaction of different policy levels — energy efficiency in buildings

The EU needs action at the local level to increase energy efficiency to mitigate climate change and to ensure energy safety. These include, low energy buildings, efficient heating systems, efficient urban design and transport. The municipalities need a supportive legislative framework and appropriate revenues from EU, national and regional levels in order to carry out these kinds of action. For instance, only the proper participation of the local level in the elaboration and implementation of the Operational Programmes of the EU Structural Funds will ensure the selection of the best local energy efficiency projects and sufficient financial means for the municipalities to implement such projects for the benefit of all.

EU level

- EU targets on climate change/energy
- Providing a regulatory framework, e.g. energy performance standards for buildings, labels for energy-efficiency of products
- Providing guidelines
- Providing Structural Funds

National/regional level

- National climate-protection programmes, targets to reduce greenhouse gases and increase energy efficiency
- Legislative framework for energy efficiency and buildings
- Energy taxes and price-support measures for energy efficiency
- Including energy-efficiency projects into the Operational Programmes of the Structural Funds

Local level

- Local targets to reduce greenhouse gases and energy consumption like the Covenant of the Mayors initiative
- Enhance energy efficiency of public buildings
- Support energy-efficiency projects by using tax revenues, Structural and other funds
- Limit urban sprawl by effective urban and spatial planning

Source: EEA, 2009.

Securing the 'right' prices

The inclusion of all environmental and social costs will ensure that consumers are encouraged to choose products and services with low environmental and social impacts. Many products sourced from remote locations may not be cheaper than local products if, for example, the environmental costs of transport or adequate labour rates in the developing countries were fully reflected in the price. At the EU level, the Greening Transport Package (COM(2008)435) includes a strategy for the internalisation of external transport costs and is thus a step in the right direction. Cities can, for instance, introduce congestion or parking charges (see also Box 2.5).

Cultures of change

Just as the car industry succeeded in associating certain cars with feelings of freedom, high status and desirable lifestyles, governments can, in addition to providing choices, support the development of a culture of sustainable choices and smart growth,

Box 2.6 Promoting cycling, walking and developing new city cultures

Copenhagen (Denmark)
Initial situation
For Copenhageners it has long been a tradition to cycle to work every day. However, in the 1960s cars started to take over more and more space in the city.

Solution
From that time the city administration dedicated more and more streets and places to pedestrians and greatly extended and upgraded the cycling infrastructure. In parallel they put a high price on parking in the city and were among the first to provide city bikes free of charge — a model copied in more and more cities across Europe. People can take their bike by public transport as well as in taxis. Many areas in the centre, like the waterfront, benefited from an attractive people-friendly urban design.

Results
All measures together supported the development of a new culture in the city — Copenhageners and their guests like to cycle, walk and meet in outdoor cafés, which was not a Scandinavian culture. Copenhagen is probably the city with the highest share of cycling to work (36 %) and walking is the dominant transport in the centre during most days. Many rankings indicate Copenhagen as one of the best cities in the world to live. The new culture strongly encourages politicians to continue this pattern of development. The boost to cycling after the introduction of free and fashionable city bikes in cities like Paris, Barcelona or Luxembourg shows the potential of cultural measures to change behaviour.

Photo: © Jens Rørbech

Photo: © Jan Gehl and Lasse Gemzøe

More information: www.kobenhavn.dk.

Skopje (Macedonia)
Initial situation
Rising car ownership became a big challenge after the fall of the iron curtain in all the eastern and south-eastern European countries. A car became a status symbol for many who consider cycling as outdated or only for people who cannot afford a car. Measures to limit car traffic and promote walking and cycling are therefore not easy to implement.

Solution
Nevertheless, Skopje is making an effort to change the trend with its transport plan. It is promoting alternative modes with an accent on bicycle riding. The city will focus on:

- improving bicycle paths;
- increasing public awareness with the Car Free Day and other campaigns;
- promoting educational events in primary and secondary school.

Since 1999, the city has participated in the annual 'Spreading Bike Riding Culture' event from March to September. So far, however, the development of infrastructure is lagging behind.

Results
Although the bicycle master plan was developed in 2003 and promotional activities and campaigns found a broad interest, the situation is still unsatisfactory: cars park on pedestrian lanes and the few bicycle lanes; the construction of cycling infrastructure is often not appropriate and is aggravated by financial problems, insufficient coordination and a changing traffic culture towards cars. As a result there is only a small increase in the number of cyclists.

This example well illustrates the need to follow a broad integrated approach and to develop the culture and supporting infrastructure in parallel. This task has proved to be difficult, in particular in cities where cycling is perceived as outdated or as something limited to activists or sportsmen, so it is hard to win support for the infrastructure. However, Skopje keeps working to solve the problems; for instance, it started to implement two pilot cycle lanes in 2008.

More information: www.skopje.gov.mk.

where quality of life is no longer defined by high material consumption (Box 2.6). As people expect change from the 'others' before changing themselves, government initiatives to encourage a change to sustainable lifestyles can have positive effects if they create the impetus and desire for change.

2.3 Urbanisation

Europe's many cities and towns, the products of a long history of urbanism and trade, are top travel destinations for tourism throughout the world and from Europe itself. Historical charm and richness, urbanity, and a diverse culture provide major tourist attractions as well as an inspiration for city planners worldwide seeking to define the compact city with human scale. However, for the vast majority of the urban population of Europe this picture of Europe's cities is very far from the reality they experience today. Urban areas are expanding more and more across Europe, increasing at a much faster rate than the growth of population. More and more city dwellers are moving outwards from the city centres into low density urban areas or the countryside but continuing to live an urban lifestyle facilitated by car-based mobility. Questions are clearly raised as to the future of Europe's cities and towns, and the quality of life that they can provide given the ever-present forces of globalisation, demographic transformations and climate change, and the impacts of many other 21st-century challenges.

This section discusses the main drivers of the urbanisation process, its various manifestations, and its effects on environment and liveability in cities. The discussion proceeds to focus on the various elements of urban governance, at different policy levels, and their impacts on the development of cities. The section aims to provide ideas and gives specific examples of the options available to minimise the negative impacts of urban living on the fundamentals of quality of life.

A Europe of cities and towns

Although, urbanisation estimates are plagued by the diversity of statistical definitions of cities and urban areas (Bretagnolle *et al.*, 2002), it still draws attention to the fact that the vast majority of Europeans — around 75 % — live in urban environments. A more comparable indicator of urbanisation in Europe is the Functional Urban Area developed in the 2005 ESPON study (ESPON, 2005a). About 1 600 settlements in Europe are considered functional urban areas, with over 50 000 inhabitants (Figure 2.10), the 75 largest and most important ones are identified as Metropolitan European Growth Areas.

Figure 2.10 Number of cities greater than 50 000 inhabitants by country

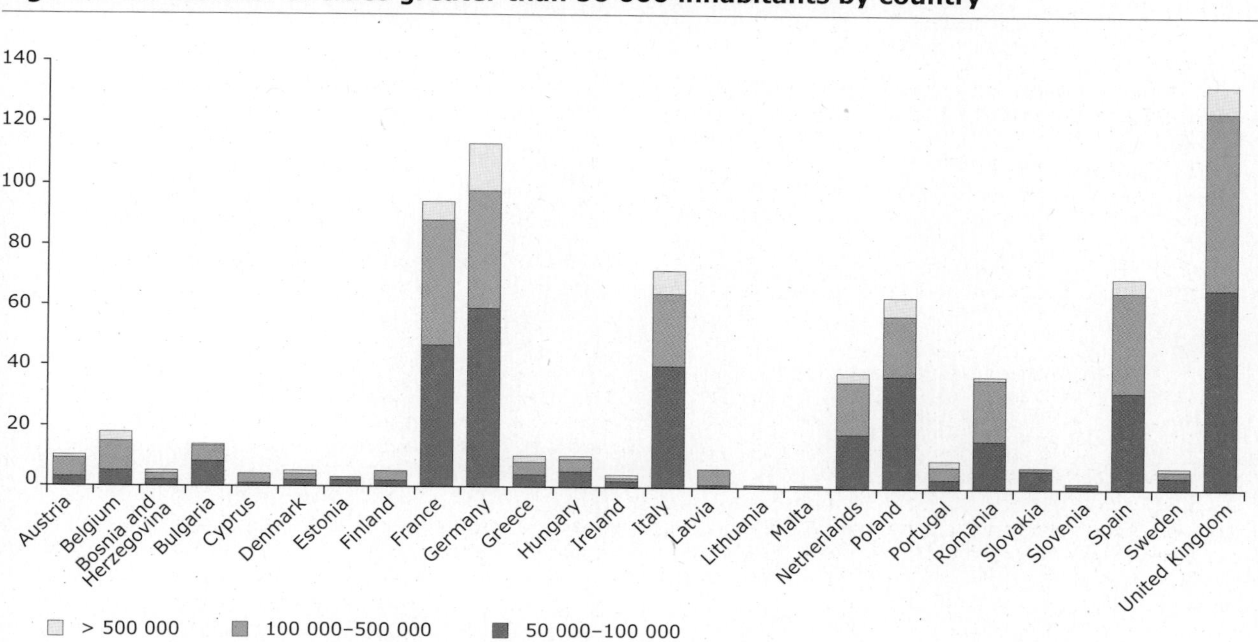

Note: Cities within each country have been differentiated according to the number of inhabitants.

Source: EEA/ETC LUSI Corine Land Cover, 2000.

Differing patterns of urbanisation

Urban land use has expanded nearly everywhere in Europe, even in areas with a declining population. Between 1990 and 2000, urban land expanded by three times the size of Luxembourg. This means an average 5.5 % increase in built-up areas, but this rate varies regionally between 0.7 % and up to 40 % This expansion is not expected to halt in the near future because it is rooted in profound long-term socio-economic changes, which deeply influence the nature, pace and pattern of urbanisation.

Urbanisation is evident in many different forms, sometimes in concentrated compact centres but typically in low density developments associated with planned or spontaneous urban sprawl (PBL, 2008 and 2009). A range of individual and only partially linked unidirectional processes influence urban systems. Consequently, there is a complex mosaic of urban growth and decline as, for instance, described in Section 2.1. Furthermore, new concepts of urbanisation have emerged, including edge cities, exurbia, peri-metropolitan areas and extended metropolitan regions (Champion & Hugo, 2004), all of which raise fundamental questions regarding the real nature and real limits of the city.

Box 2.7 shows these different urbanisation patterns across Europe. Some regions are experiencing compact forms of urbanisation, often accompanied by rapid population growth, but urbanisation patterns are mainly characterised by rapid urban growth and decreasing population densities in residential areas. Accordingly, the general trend has been the expansion of urban land use, with fewer people inhabiting more urban land and thus contributing to urban sprawl.

Multiple drivers of urbanisation

Urban change can be defined in terms of five main components or domains: society, economy, built environment, natural environment and governance. Figure 2.11 shows the main factors driving changes in these domains. Economy and

Figure 2.11 Main European drivers of urban change

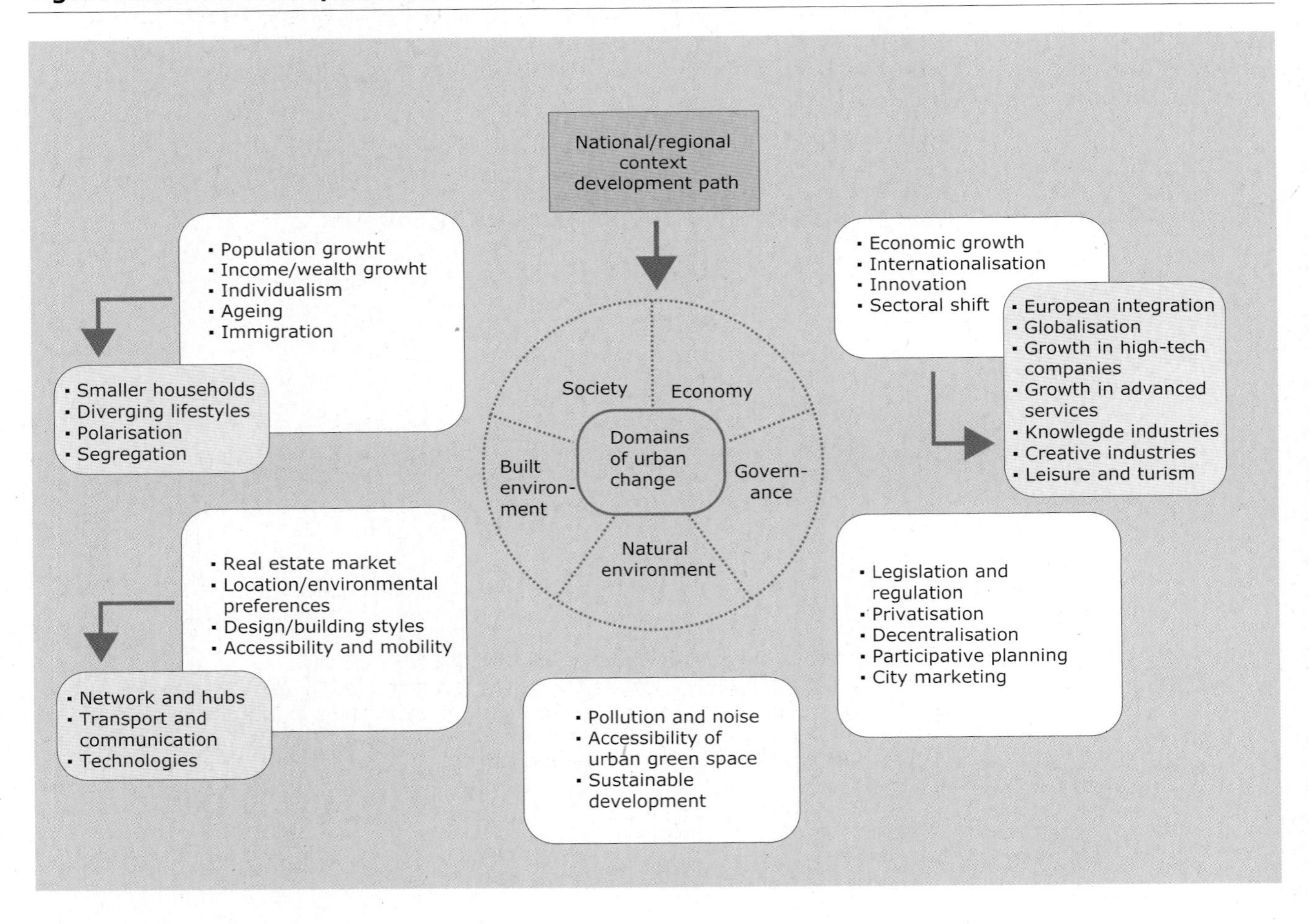

Source: PBL, 2008.

Box 2.7 Urbanisation in European regions

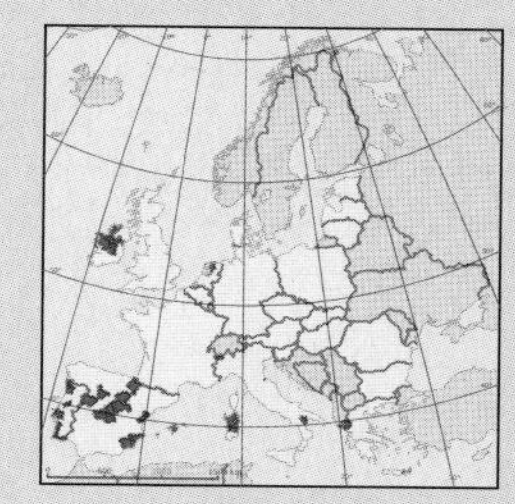

Very rapid urbanisation
Regions of this type, which experienced the largest population growth in Europe, can be found along the Portuguese coastline, in Madrid and its surroundings as well as in some coastal regions in Spain, in the north of the Netherlands, and north-western Ireland. Some regions in Italy (especially the north of Sardinia) and Greece also belong to this group. Urban land cover has been increasing between four to six times faster than the European average, but the population density in residential areas declined six times faster than the European average. This suggests a pattern of low-density housing construction.

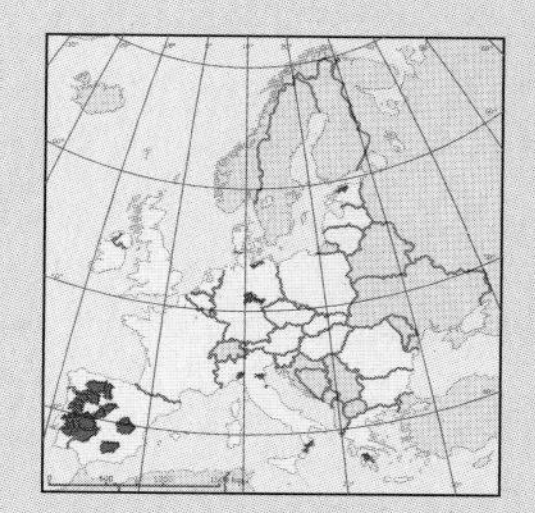

Rapid urbanisation with a declining population
Regions on the Iberian peninsula around large cities such as Madrid, Porto and Lisbon are typical of this type of urbanisation, as are a few regions in Italy, Erfurt and Rostock in Germany, Tallinn in Estonia and Arcadia in Greece. The urban land cover in regions in this category increased about three to five times faster than the European average. As in regions which have experienced very rapid urbanisation, the population density in residential areas fell, six times faster than the European average. But, unlike the former category, the total population also fell by an average of 0.6 % per year. Urbanisation here demonstrates not only population decline, but at the same time continued peripheral growth of the built-up area, highlighting the power of the forces of sprawl.

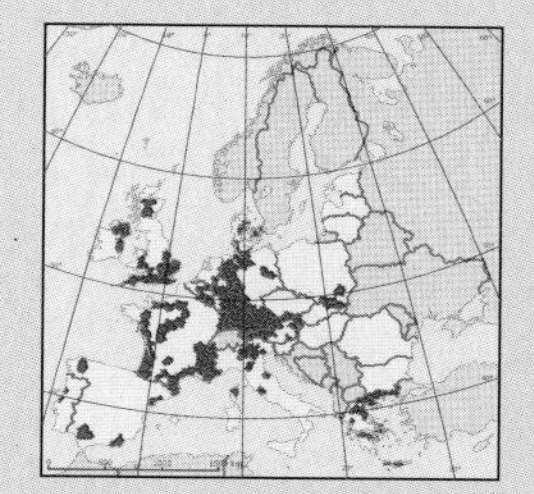

Compact urbanisation with rapid population growth
This category can be found all across Europe: in the western parts of Germany, in Paris and the coastal regions of France and Spain, Austria, northern Italy, Greece, southern UK, Scotland, the east of Ireland, and some regions in the south of Poland and Hungary. The regions of this cluster are characterised by an average increase in built-up land cover; however, population increased very rapidly, the most rapid growth in all types of urbanisation. Consequently, the population growth in residential areas increased most rapidly as well. This densification suggests high-rise inner city construction as well as increases in inner city populations.

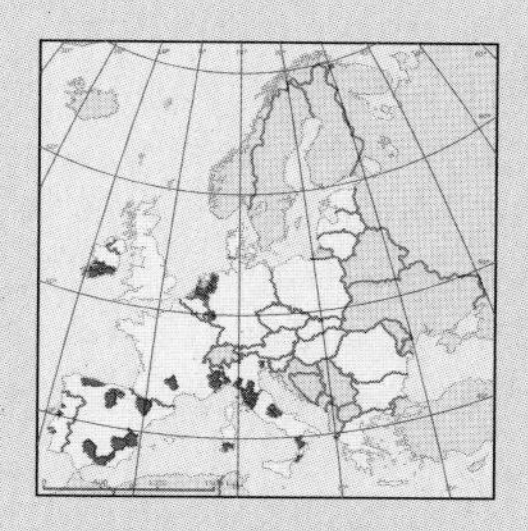

Rapid urbanisation but low density
Most regions in the Netherlands belong to this category, as well as southern Ireland and a number of regions in Spain, Portugal and Italy, mainly on the coast. The regions in this group are characterised by rapid urban growth, which is two to three times faster than the European average. Nevertheless, the population density in residential areas has declined relatively fast, probably due to urban expansion with relatively low density. Regions with very rapid urbanisation have the same characteristics, though development takes place at a faster pace.

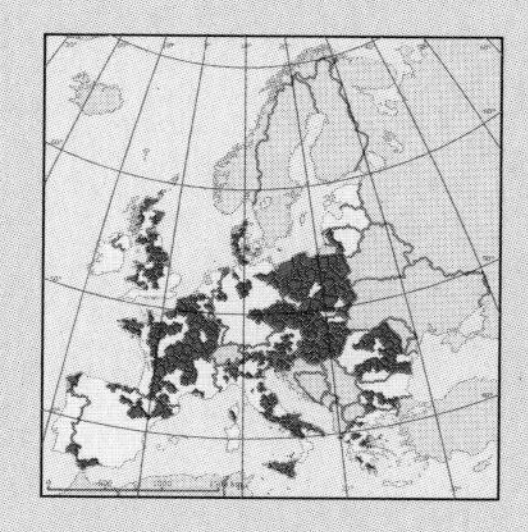

Slow urbanisation
This form of urbanisation is characteristic of many countries in Eastern Europe as well as in peripheral regions in almost all other European countries, most notably in the United Kingdom, Denmark, Belgium, France, parts of Austria and Italy. There is a relatively slow increase in urban land cover and population growth, both about one third of the European average. The population density in residential areas has declined slowly, about half as fast as the European average.

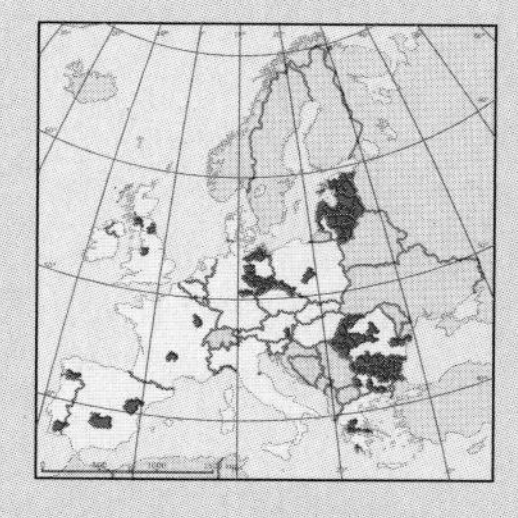

Slow urbanisation with a declining population
This type of urbanisation is characteristic of most regions in the Baltic States and eastern Germany, as well as regions in Romania, Bulgaria and isolated regions in the United Kingdom, France, Austria, Italy and on the Iberian Peninsula. The urban land cover in this category grows particularly slowly, accompanied by a rapid decline in population. Unlike other clusters, the population migrates towards other regions, leading to lower population densities in residential areas.

Source: PBL, 2008.

changes in demography, like ageing, migration or the trend to smaller and more households with new demands for housing and land consumption (see in more detail Section 2.1) compete as prime, often interrelated drivers. In some parts of Europe, population growth is an important driver of urban growth, for example Madrid; but more important is the enormous growth of the number of (small) households as a result of ageing and broad individualisation trends, such as the increase in the number of divorces and young people leaving the parental home at an earlier age. In addition, economic development attracts population by domestic and international migration. However, whilst the environmental components seem to be less significant in the short term, they tend to become increasingly important in the longer term as preconditions for attractive and healthy places to live and work. Finally, governance is both a facilitating and a steering driving force.

Shifts to service economy

Over the past decades, major changes in the EU economy have occurred due to globalisation. In combination with technological change, rationalisation, and European integration, this has led to the transfer of much of the production of capital and consumption goods to regions both within and outside Europe that can offer cheaper labour (Dicken, 2004; Eurostat, 2007). Consequently, many European urban economies have made a further shift towards service-oriented urban economies.

Figure 2.12 Built-up area, road network and population increases

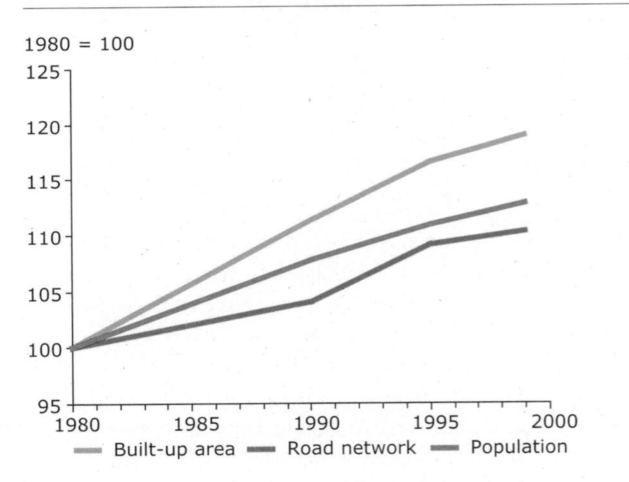

Note: Selected EEA countries: Belgium, Czech Republic, Denmark, France, Germany, Latvia, Lithuania, the Netherlands, Poland, Slovakia and Spain.

Source: EEA, 2006.

Photo: © Google Earth

Business services, including financial services, form the largest economic sector in the EU, and accounted for more than a quarter of the EU-25's gross value added in 2005 (Eurostat, 2007). These changes in urban economic structures have had, and still have, a major impact on urbanisation through the creation and loss of employment opportunities and economic growth in cities. A flourishing economy creates a demand for offices, industrial buildings, houses and other activities to accommodate the population and business sector, and will also support more and better public facilities. Conversely, urban areas offer agglomeration advantages, which attract and can stimulate further economic activities (Porter, 1990). Economic stagnation has reverse effects. The economic situation of urban regions, and income distribution within them, are therefore important determinants of both objective and subjective views of quality of life and place.

New mobility stimulates suburbanisation

The steady growth of net income during the last decades and the increased share of earnings that could be spent on other than basic needs meant that people could afford bigger and remote places to live. Building styles, types of house and neighbourhood lay-outs have varied throughout Europe and in time, depending upon economic situation, housing culture, and population composition. Nevertheless, residential preferences have typically shifted towards low density housing in greener environments.

The realisation of these low density urban expansions — urban sprawl — has been helped along by the enormous rise in car ownership that began in the 1960s. Rising net income and mass car ownership have resulted in continuous waves of suburbanisation throughout Europe, fluctuating according to the waves of urban growth and expansion of the road networks. The construction of new infrastructure, in particular motorways — the total length of the motorway network has tripled in Europe over the last 30 years — has substantially enhanced the accessibility and thus the attractiveness of specific localities for single family houses, second homes and business. A good example is the Compostela–La Coruña–Porto–Lisbon motorway, which follows the coast of the western Iberian Peninsula. Many new urban and business areas have developed along its length. It has considerably improved accessibility in a north-south direction, making this axis highly attractive to many companies, facilitating economic growth in the region, and forming a new market for labour force recruitment (Lois-González, 2004).

Similarly, developments in air and rail transport have also stimulated further urbanisation and influenced the shape of urbanisation. Between Paris and Brussels urban growth has occurred around 'beetroot' stations on the TGV railway (EEA, 2006b).

For economic reasons, investors target greenfield-sites for new developments. Agricultural land allocated for development is significantly cheaper than the expensive land and apartments in core cities, dilapidated residential areas and brownfield sites.

Compromising life-support systems

Urban expansion is often perceived as a route to a better quality of life as it offers affordable, greener places to live. But related transport infrastructure developments may lead to further deterioration and fragmentation of natural areas and valuable landscapes, thus resulting in a less biodiversity and the deterioration and loss of ecosystem services — flood prevention, water clean-up, climate regulation etc. (EEA, 2006a). Land take may also reduce the area available for food production. The increasing pressures on the food market due to growing worldwide demand, and competition with bio-mass production related to high fuel prices, may heighten the importance of the loss of agricultural areas in the future. All this impacts the quality of life of all Europeans dependent on these basic life supporting services now and in the future.

In addition, urban expansion goes hand in hand with a range of problems concerning water quality and quantity. There may not be enough clean tap water locally to meet demand and urban water usage may have induced salinisation in coastal regions (BPL, 2005). In some areas with great pressures for urban growth, construction has taken place in areas that are vulnerable to flooding (Sagris *et al.*, 2006; Barredo *et al.*, 2005). Also, building on the flood plain, increases the risk and severity of flash floods (see Box 2.8).

Sprawl triggers transport growth

Many problems in cities are strongly related to issues concerning urban density and urban containment. Lower residential densities often offer lower noise levels, less air pollution and better access to (private) green space. On the other hand, low densities also result in greater demands on the transport system, particularly road transport. Hence urban sprawl and transport infrastructure have a reciprocal relationship and a positive feedback loop develops (ESPON, 2004) — more building requires more roads, which leads to more building (see simulation in Box 2.9). Transport volumes have increased substantially throughout Europe over the last decades driven by urban sprawl and a large number of other socio-economic factors (Stead & Marshall, 2001).

Growth in transport demand has increased the emission of greenhouse gases from urban areas, and has exacerbated the problems of climate change (Section 2.5). Many cities, especially the larger ones, also suffer congestion on their roads and the amount of space that is set aside for roads means that there is a lack of public space for leisure activities, walking and cycling. Transport-related noise and air pollution increase health risks and reduces the quality of life in cities (Sections 1.2 and 2.4).

Challenges of compact cities

In contrast to the general lowering of urban densities, some cities experience growth in the inner city, which results in areas of high population densities. On the positive side this generates the potential to reduce transport demand and overall emissions, but on the negative side there is a risk that more people are exposed to higher levels of air pollution and noise. Urban design, spatial planning and other administrative measures can reduce these impacts to some extent, as has been demonstrated by the revitalisation of many inner cities in the 1980s and 1990s. However, unfavourable living conditions in inner city areas associated with excessively high population

Box 2.8 Dresden-Prague corridor (Germany, Czech Republic) — urban expansion and the impact of flooding

The German reunification and the collapse of the communist block led to major changes in economic regime from a planned to a market economy in both the former East Germany and the Czech Republic. These changes have created completely new driving forces for urban development in the transport corridor between Dresden and Prague. As a result, the built up area has grown substantially since 1990 (EEA, 2006a).

The corridor mainly follows the Elbe river that is plagued by major flood events, such as the dramatic flooding in 2002 which had high human and economic costs. The analysis below shows that the vulnerability of urban areas has increased and the effects are at least partially man made.

Map 2.4 shows the land-use and flood hazards in the Elbe river catchment area. The changes in the exposure to floods during the period 1990–2000, which is given by the total surface (in km^2) in flood-prone areas, indicate an increase of urban areas of about 50 km^2 (Figure 2.13). That means that many new residential areas have been built in flood-prone areas and are more vulnerable now.

Map 2.4 Elbe catchment area: the Dresden-Prague corridor

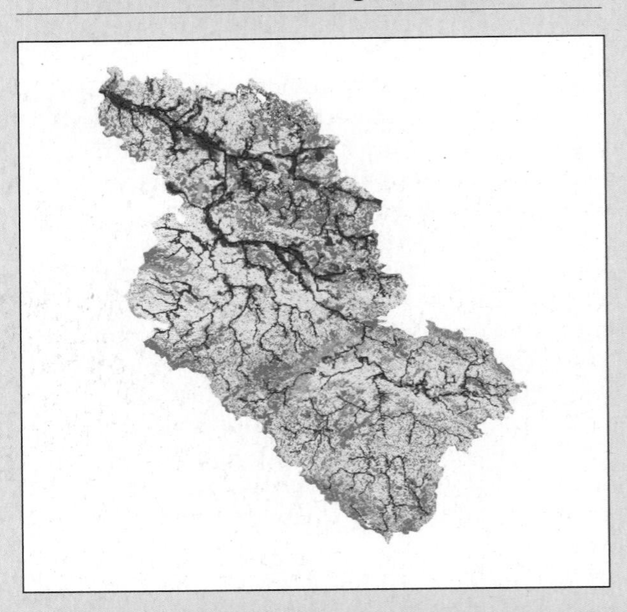

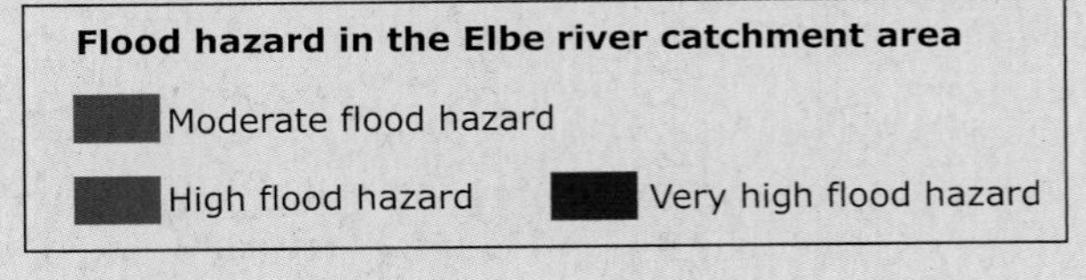

Figure 2.13 Elbe catchment area: evolution of exposure to flood in the period 1990–2000

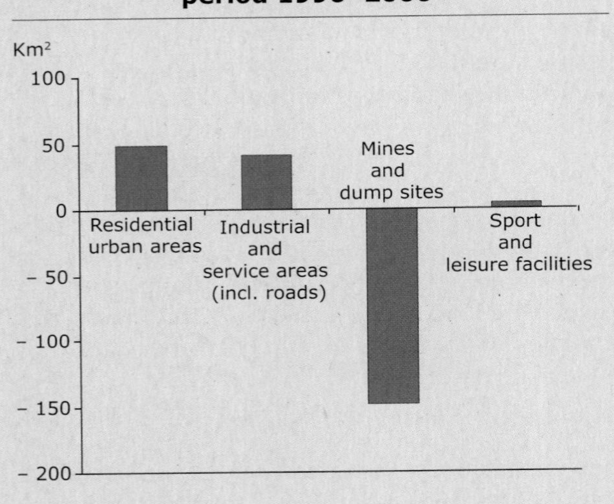

In different projections (Figure 2.14), even more built-up areas are expected to become vulnerable. Commercial areas are projected to be the most exposed to floods in all scenarios. So far no prevention actions have been considered in the simulations.

Box 2.8 Dresden-Prague corridor (Germany, Czech Republic) — urban expansion and the impact of flooding (cont.)

Projections

To model future impacts three different development scenarios were simulated for the Dresden-Prague corridor:

- Business-as-usual: extrapolates moderate 1990s trends of land-use change, indicating that the land-use patterns of the area will not change considerably over the next two decades.
- Built-up expansion: elaborates the socioeconomic projections of the European Environmental Agency.
- Motorway impact: evaluates the impact of motorway development (A17/D8 part of TEN Corridor IV).

Figure 2.14 Projected estimate of exposure to flood for artificial land-use classes

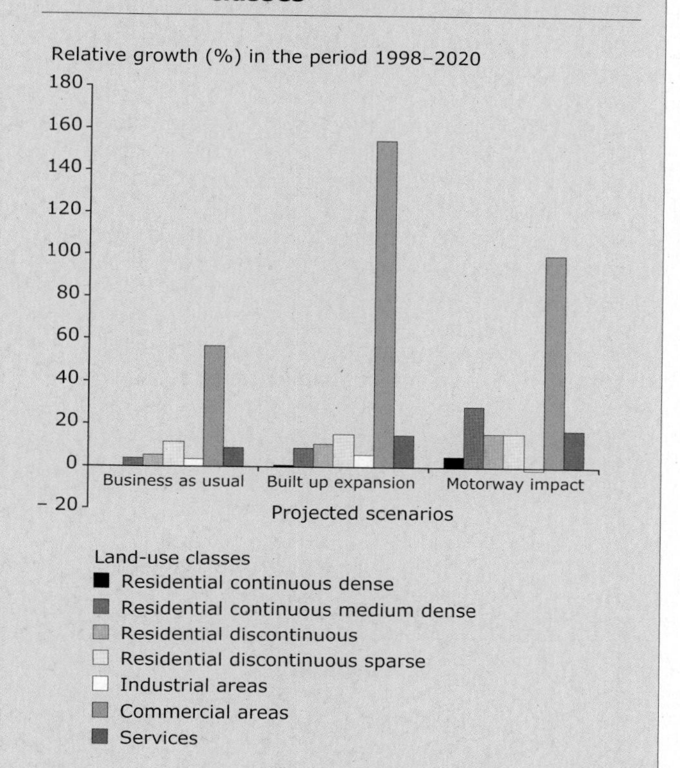

Source: JRC, 2009.

densities can also contribute to suburbanisation and ex-urbanisation and thus reinforce the tendency to urban sprawl.

A further problem of urban living is the lack of green areas in many highly urbanised regions. Urban expansion and higher densities have often led to growing separation of home and recreational areas, and the reduction of parks and playing fields, limiting the possibilities for outdoor recreation. Finally, the deterioration of landscapes and natural areas surrounding the cities as the low density urban expansion impinges on the countryside is associated with adverse impacts on social life, physical activity and mental health (see also Section 1.2)

Social inequalities

Urban growth also affects the spatial organisation of cities. Typically, suburbanisation and urban sprawl have promoted segregation and polarisation along ethnic or socio-economic lines (see PBL, 2008 and 2009). The real estate market plays a central role in these intra-urban developments, resulting in a qualitative spatial sorting of employment sectors, population groups and public space. Lower income households cannot afford homes in high price areas, and usually live in areas of dense housing with less green and good quality public space, higher noise and air pollution levels or far away from attractive urban areas. These segregation trends lead to temporary and more permanent unequal developments, loss of social balance and cohesion. The resulting imbalances show themselves socio-economically in the exclusion of specific groups from employment and services like culture and education, and by accumulation of socio-economic and environmental problems in deprived areas. Other factors exacerbate the situation: growing urbanisation is accompanied by

Box 2.9 Scenario for the Dresden-Prague corridor — transport growth driven by increasing urbanisation

As described in Box 2.8 East Germany and the Czech Republic underwent major economic changes, creating new driving forces for urban development after the collapse of the communist block. Moreover, EU membership has led to a growing engagement with European markets and access to EU development schemes e.g. TEN-T, Structural Fund. Scenarios show that the share of urban areas is expected to grow in every case, although the amount differs depending on the assumed development path: business-as-usual, the strong expansion of built up areas or the impact of the motorway development (A17/D8 part of TEN Corridor IV) (EEA, 2006).

Map 2.6 shows the modelled transport growth caused by new residential and business areas in the Dresden-Prague corridor (Map 2.7), which is substantial in many areas. The increase of traffic intensity is particularly evident in the centres of Dresden and Prague.

Map 2.5 Dresden-Prague corridor, scenario location

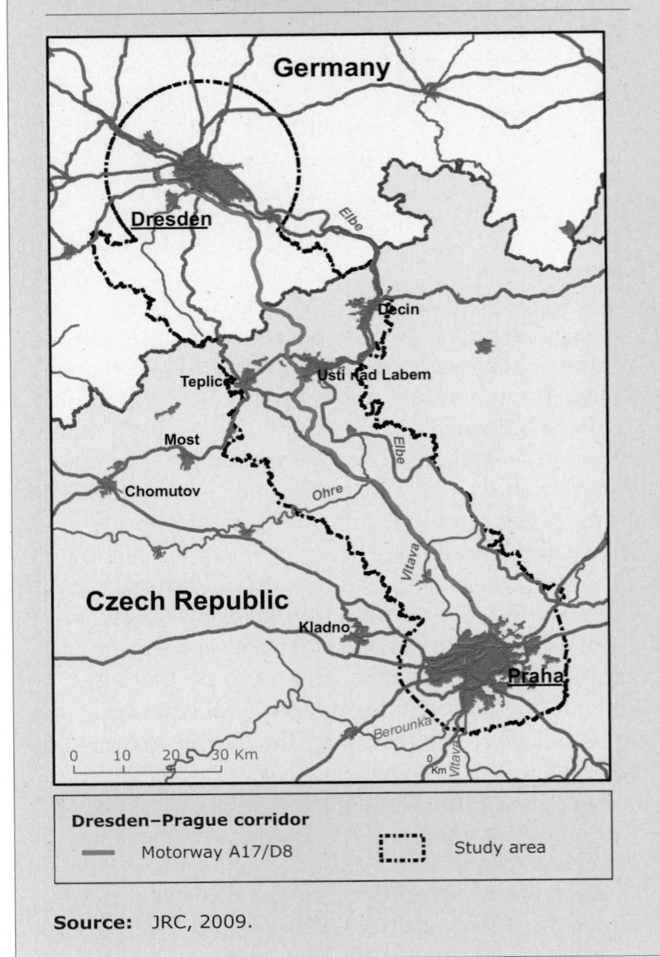

Map 2.6 Max road-traffic intensity (no of cars/hour)

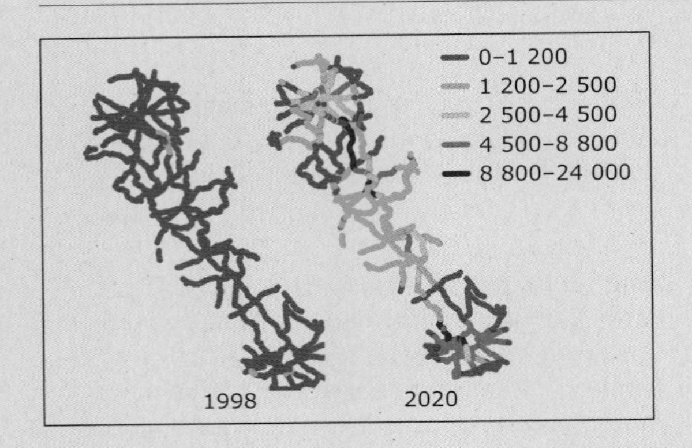

Map 2.7 Development of Prague (2020) according to the EEA Outlook scenario

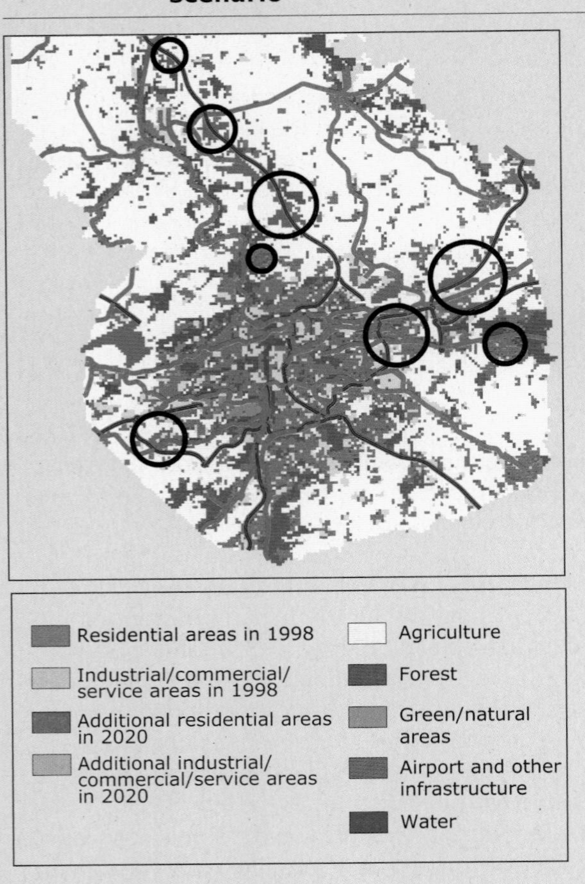

Source: JRC, 2009.

an increasing share of young adolescents, ethnic and socio-economic segregation, increasing crime and reduced levels of personal safety (Dij, 1999; Hideg and Manchin, 2007).

Planning sustainable cities

The extent to which urbanisation is managed depends upon the planning system in each Member State. It is predominantly the domain of local and regional governments, and in some countries also that of national government. A range of instruments, including land and housing prices, spatial planning, urban design, building regulations, taxation and urban planning, play a central role providing the basis for an integrated spatial approach. A large variety of urban planning systems and cultures are evident throughout Europe; however, their effectiveness depends on the level of liability, their scope, the planning culture and history, and also on the extent of influence of national governments on spatial planning (Haskoning, 2008).

Some cities and towns have started to develop more comprehensive strategies for sustainable development (Box 2.10). In the context of the pioneering work of the Aalborg Charter ([5]) and the Sustainable Cities and Towns Campaign ([6]) the aim is to secure the integration of social, economic, and environmental dimensions. An important element of this work is the exchange of good practice between municipalities maintained via the supportive activities of the various pan-European city networks. Many comprehensive rehabilitation measures in cities and towns over the last decades, such as pedestrian zones, redevelopment of brownfield sites, public places and green areas of high urban quality, and affordable housing, have contributed to making city centres more attractive as places to live.

Supportive European policy

Spatial planning is not a formal competence of the EU; nonetheless, the allocation of Structural Funds, the EU Transport Policy and other policies have a big impact in stimulating and restructuring existing urban areas and supporting the development of new urban centres.

The 1999 European Spatial Development Perspective (ESDP), a non-binding framework that aims to coordinate various European regional policy impacts, was a result of a Member States initiative during the 1990s. The document advocates the development of a polycentric and balanced urban system and strengthening of the partnership between urban and rural areas; parity of access to infrastructure and knowledge; and wise management of natural areas and the cultural heritage.

Recently, the *Green Paper on territorial cohesion* (EC, 2008c), the 2007 EU Territorial Agenda and *Leipzig Charter on sustainable European cities* built further on the ESDP. Today the leading theme of regional and urban-oriented policies at the European level is *cohesion*, which aims to promote socio-economic convergence and coherence among and between the regions, and in the cities of the union, thus ensuring a high quality of life. The European Commission supported many urban projects via the Structural Funds and will strengthen its support for urban rehabilitation over the next funding period (see more in Section 2.6).

The EU Transport Policy promotes effective and sustainable transport systems. The realisation of the Trans European Transport Networks (TEN-T) aims to create improved accessibility for all cities and regions in Europe (Ravesteyn and Evers, 2004; High Level Group, 2003). Other EU policy provides guidance on the development of more sustainable urban areas, in particular, the *Thematic Strategy on the urban environment* (EC, 2006d) and the Green Paper *Towards new culture for urban mobility* (EC, 2007d). Furthermore, several EU programmes promote and fund sustainable urban development, including the LIFE+ programme for the environment, the EU Seventh Research Framework Programme, and the CIVITAS Initiative for clean and better transport (see also synthetic Table 1.2). Competitions like the European Green Capital Award stimulate more action.

Barriers to effective policy-making

Notwithstanding all these positive approaches, Europe and its cities and towns must still meet the challenges of unsustainable urbanisation patterns, including ongoing urban sprawl, and there are still important gaps in current policy-making.

European policy also influences urbanisation patterns indirectly, and these indirect influences can be both supportive and inhibiting of sustainable urban development. For example, as a result of

([5]) http://www.aalborgplus10.dk/default.aspx.
([6]) http://www.sustainable-cities.eu/.

Box 2.10 Bologna master plan (Italy) — integration of energy policies

Context of Municipality and initial situation
Bologna is a city of 390 000 inhabitants at the centre of a larger metropolitan area. Its economy is based mainly on knowledge (university) and mechanical industries. The city is situated at the heart of freight and passenger traffic transiting between the Mediterranean area and other European regions.

During the last 15 years, the city's CO_2 emissions have been constantly increasing at a yearly rate of about 1.3 %. Energy consumption data show that the housing sector is responsible for about 62 % of overall emissions and is caused by an average efficiency of existing buildings that is far below modern standards.

Photo: © Daniele Zappi

The case
In order to introduce policies for a substantial emission reduction, Bologna developed the new City Energy Programme (2007) with the goal of a 28 % reduction of greenhouse gases emissions. Its energy-saving measures and promotion of renewable energy sources are based on the close integration of an analysis of energy issues and the development of appropriate urban planning tools. The underlying CO_2 emission analysis is based on bottom-up reconstructions, and considers the census figures and consumption by individual buildings using available GIS databases. The data are collected in the Energy Atlas, and the resulting geographical platform allows for the:

- identification of urban areas with the highest energy intensity;
- identification of specific areas and buildings that may be the object of direct improvement;
- assessment of the energy-related environmental impacts of new urban developments.

When developing the broader Municipal Structural Plan, the city integrated these results. The plan was approved in July 2008 after a complex process, including public and institutional participation, and sets the principles that will guide the development of the city over the next fifteen years. By using the energy analysis, areas affected by highly significant urban transformation have been organised in clusters, called Urban Energy Basins. They form homogeneous areas in which the city applies specific energy policies. The 11 energy basins identified cover about 15 % of the city's territory. The energy impact of predicted transformations has been calculated in detail for each area.

The Municipal Structural Plan provides for specific sustainability measures in any urban development area, enabling lower energy consumption and sustainable energy supply by recommending appropriate urban population densities and reductions of transport demand in particular. A set of building rules provides a comprehensive technical tool for professionals working on city development projects to integrate energy savings measures. The energy standards contained in these rules have been set at even higher levels in the Urban Energy Basins according to their characteristics.

Box 2.10 Bologna master plan (Italy) — integration of energy policies (cont.)

Results, lessons learned and transfer potential
The energy standards will affect both new constructions as well as the rehabilitation of existing buildings, achieving a 20 % reduction of CO_2 emissions in the housing sector in 15 years. The integration among different urban planning instruments is generally replicable because it is based on commonly available GIS tools.

The broader integrated approach allowed for the inclusion not only of measures to reduce the energy consumption of buildings but also measures in other professional areas such as urban design to further reduce Bologna's overall energy consumption. This work won the first prize in the Sustainable Energy in Cities in the 2008 contest promoted by the Italian Ministry for the Environment and the National Institute of Urban Planning within the Sustainable Energy Europe campaign (SEE).

Source: New city master plan (PSC) http://www.comune.bologna.it/psc, Energy Programme (PEC). http://www.comune.bologna.it/ambiente/QualitaAmbientale/Energia/PEC/Programma.php. Sustainable Energy in Cities award promoted within SEE campaign. http://www.campagnaseeitalia.it/news/concorso-energia-sostenibile-nelle-citta.

Box 2.11 Fragmented decision making — motivations of actors involved in land development

Municipalities maintain the hope that new inhabitants will lead to a tax surplus, when in fact studies have shown that this is seldom the case. Therefore they generally favour the development of land. Costs are transferred as far as possible to the investor and as the municipality bears no costs the project is regarded as beneficial.

For landowners a plot represents an economic asset in whose increasing value they hope to profit. Therefore, owners of agricultural land with prospects for development become highly active.

For project developers high unit costs to connect new dwellings or commercial premises to supply networks are often more than offset by the much cheaper land prices in areas at the edge of existing settlements. The extra transport costs are countered by other sales arguments (e.g. property prices, 'living in the countryside').

Utility companies have little motivation to influence the location and density of use of newly constructed or newly connected areas as the associated costs are reimbursed by users in the form of construction subsidies or by a general raising of charges for all users.

Householders seeking a new location are often ignorant of the high costs for technical infrastructures associated with low density peripheral areas. The low price of suburban land hides the rising infrastructure costs per housing unit which is caused by low settlement densities.

Mixed motivations of actors in the development of land supports fragmented decision-making and unsustainable land-use development: the individual decisions are rational, but when actors ignore the high follow up cost for transport, infrastructure, loss of land, biodiversity and ecosystem services they transfer these costs to others and eventually to every resident.

Source: UBA, 2009.

EU Transport and Common Agricultural Policy, some rural areas have become more accessible, which has encouraged urban sprawl and hence increases in commuting. So far, these indirect impacts have not assessed.

Given the significant interlinkages between local development and the impacts of European policy, it is clear that a narrow interpretation of the subsidiarity principle ([7]), restricting effective engagement of the EU at the local level, is inhibiting the development of solutions to urban problems. Many problems, including those of urban sprawl, definitely have a European dimension, which requires adequate consideration.

Furthermore, practical decision-making is often fragmented. Typically individual decisions are supportable, but taken together they can lead to contrary effects and unsustainable urban development, as evident in the example of decisions related to land use (Box 2.11).

Spatial planning, if used properly, can be an appropriate way to balance disparate demands on land use. However, traditional top-down spatial planning, in the form of zoning, is not in itself sufficient to steer urbanisation. Such planning is unable to manage, for example, effectively those drivers that cannot be expressed in spatial terms, it does not engage effectively with 'bottom up' and participation processes, and is relatively static and not responding sufficiently rapidly to the dynamic of many urban processes. It needs to be complemented with other instruments and be part of a much broader management approach.

Overcoming barriers to action

Despite the many problems seen today, urbanisation is manageable, and more sustainable development should always aim to provide opportunities for more efficient urban living and quality of life.

European policy impacts

The European Commission, to counter the different intended and unintended impacts of European policies on urbanisation, needs to cross-check for negative impacts in all its policy areas and aim to find integrated solutions. The new approaches embodied in the territorial cohesion initiatives may provide the major impulse needed to support the development of such an integrated cross-thematic policy framework, supported by a comprehensive knowledge base about the potential and real impacts of EU policies.

Horizontal and vertical integration

Given the many drivers that influence urban development and urbanisation, it is clear that the issue of policy integration is a key concern. The need for horizontal integration is widely recognised, but effective implementation remains problematic in many cases, as the example in Box 2.11 demonstrates. In particular, the sustainable development strategies and action plans provide a good basis for integration, but they need to be backed by stronger political commitments, the real participation of all relevant stakeholders and institutional integration.

Problems associated with urban sprawl and climate change cannot be solved at the local level alone. The principle of subsidiarity needs to be reinterpreted so that administrations recognise that urbanisation problems require active participation at various administrative levels in order to generate integrated policies.

Complementing traditional spatial planning

The management of urbanisation also needs to influence non-spatial issues. Thus, spatial planning needs to become part of a broader governance approach that includes participation, mediation and changing lifestyles, and balances the economic, environmental, social and cultural aspects (see example in Box 2.12). Beyond plan zoning, the development of economic tools is necessary to close the price gap between agricultural and urban land.

Increasing urban attractiveness

Making cities more attractive by enhancing the factors — social, cultural, economic and environmental — that contribute to quality of life — stimulates people to live in the cities themselves, which keeps cities compact and avoids urban sprawl. Local measures supported by regional, national and European policies include safe and usable public places of high aesthetic quality, green areas and corridors, low noise and air pollution levels, good quality and affordable housing, integration of immigrants and other social groups,

([7]) The subsidiarity principle requires that matters in the EU are handled by the lowest competent administrative level.

Box 2.12 Participatory land-use planning in Freiburg (Germany)

Initial situation and context of the municipality
In spring 2001, the municipal council of Freiburg, a city situated in the South-west Germany, between the Black Forest region, Switzerland and the Alsace region, decided to review the land-use plan for 2010, paying particular attention to the needs of citizens. This attention was due to an outcome of previous participatory processes that had failed, resulting in growing mistrust of citizens in the transparency of the government. The result was a clear call for more active public participation from the very start.

Photo: © Holger Robrecht, ICLEI

Solution
To engage the public, the council established a systematic process in which citizens were involved in the decision-making processes and thus able to actively participate in the development of Freiburg's land-use plan. To allow for maximum input, the land-use plan was extended up to the year 2020 and a cross-departmental project steering group for integrated urban development was set up in the Mayor's office.

In the two years that followed, Freiburg succeeded in continuously integrating citizens in the development of Freiburg's land-use plan, thereby eliminating any mistrust. Ongoing public participation in the land-use plan was organised into three stages:

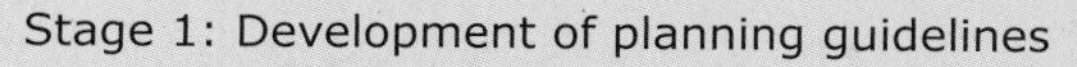

Stage 1: Development of planning guidelines

Stage 2: Information events

Stage 3: Facilitation of working groups and extended civil participation

Results, lessons learnt and transfer potential
The citizens' opinions about the land-use proposal were introduced to the public and passed on to the mayor as a 'vote' by Freiburg's citizens. Consequently, the administration changed the draft land-use plan and the municipal council approved most parts of the citizens' land-use proposals. The number of citizens involved and their ongoing commitment to the process was notable.

This process has shown the value of an accurate review of the issues and roles in the preparatory phase. The administration was able to realistically outline the planning process, thereby anticipating potential conflicts and allowing for solutions to be found. Furthermore, external facilitation helped in establishing the concept of the 'Freiburg model' and supported mediation between the citizens and the administration in conflict situations. In addition, the involvement of competent citizens as voluntary facilitators led to many positive results in the short amount of time given. From all angles — participants, citizens, the administration, and the municipal council — the process of public participation is rated as a success and Freiburg plans to apply it in planning processes in the future.

Source: Bürgerbeteiligung zum Flächennutzungsplan 2020, www.stadt.freiburg.de/1/1/121/index.php.

jobs, mobility, a rich cultural life and a positive sense of identity.

2.4 Air pollution and noise

Sounds such as a train in the distance, church bells, the shouting of market traders and children playing are, generally speaking, accepted as a welcome fact of life. All these sounds, the natural, social and urban sounds, connect you with the world around you and contribute to quality of life. But what if the soundscape is dominated by traffic noise, the noise of aircraft and nearby trains?

Despite past measures and many improvements, noise and air pollution in many Europe's cities is still high and above healthy limits, leading to various types of disease and shortening life expectancy. These challenges for human health, the environment, and finally people's quality of life are very complex and must be tackled at every administrative level. This section gives an overview of the problems and suggests how they should be managed and the action required at the local, regional, national and European levels.

Air pollution is still a serious threat

Across Europe, people are exposed to levels of air pollution that exceed air quality standards set by the EU and the World Health Organization (WHO). This occurs mainly within urban/suburban areas (Figure 2.15). In the period 1997–2006, 18–50 % of the urban population was potentially exposed to ambient air concentrations of PM_{10} higher than the EU limit value set for the protection of human health. There was no discernible trend over this period (EEA, 2007a).

For ozone (O_3) there was considerable variation over the years. During most years, 14–61 % of the urban population was exposed to concentrations above the target value. In 2003, a year with extremely high ozone concentrations due to specific meteorological conditions, the exposure to high concentrations increased to about 60 %.

In the period 1997–2006, 18–42 % of the urban population was potentially exposed to ambient air nitrogen dioxid (NO_2) concentreations above the EU limit value. The percentage of the urban population exposed to SO_2 concentrations above the short-term limit values decreased to less than 1 % and the EU limit value is thus close to being met (EEA, 2007a).

Many European urban areas experience daily average PM_{10} concentrations higher than 50µg/m^3 on more than the permitted 35 days per year (Map 2.8). The highest urban concentrations were observed in cities in northern Italy (Po valley), Spain, Portugal, the Czech Republic, Poland, Hungary, Romania, Bulgaria, the Benelux countries, Greece, and the cities of the West Balkan countries.

As a result, the exceedance of air quality standards seriously increased respiratory and cardiovascular diseases, in particular in young children and the elderly. There seems to be a strong relationship between the amount of heavy traffic and the health effects; epidemiological studies in the Netherlands, for instance, show a greater incidence of respiratory and cardiac disease in people living near major roads (Hoek *et al.*, 2002). In the European Union,

Figure 2.15 Percentage of the urban population in EEA member countries (except Turkey) exposed to air pollution above the limit and target values

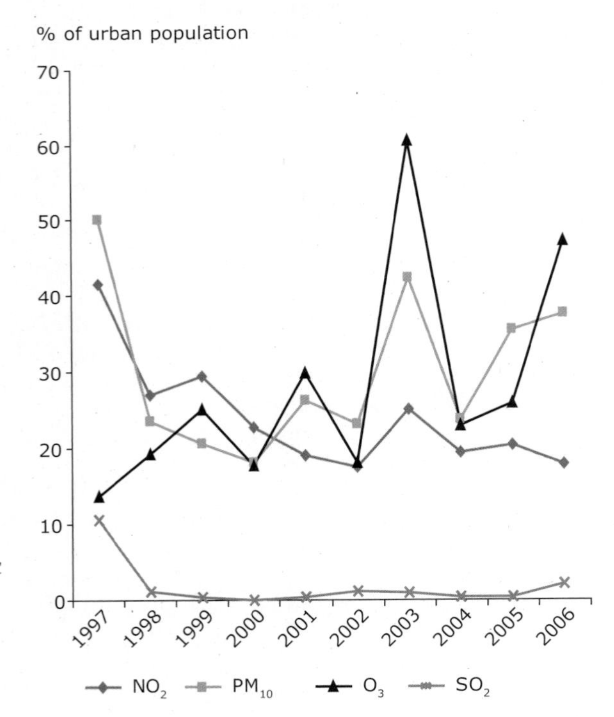

Note. Limit values are: PM_{10} — 50 µg/m^3 24-hour average not to be exceeded for more than 35 days; NO_2 — 40 µg/m^3 annual average; SO_2 — 125 µg/m^3 24-hour average not to be exceeded for more than four days; O_3 —120 µg/m^3 8-hour daily maximum not to be exceeded for more than 25 days averaged over three years.

Source: AirBase.

Map 2.8 **PM_{10} showing the 36th highest daily values at urban background sites superimposed on rural background concentrations, 2005**

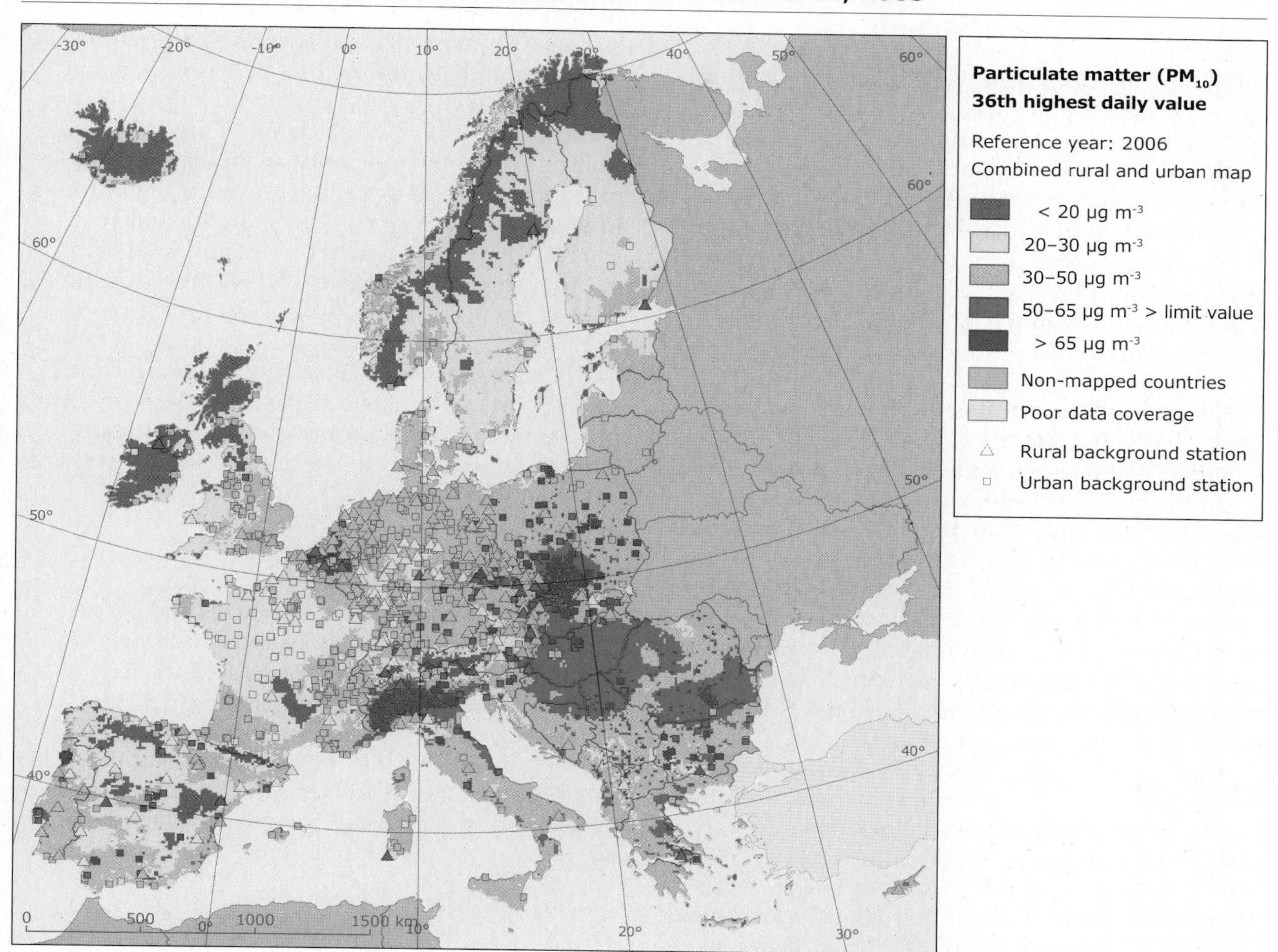

Source: EEA, AirBase.

the number of premature deaths that can be attributed to anthropogenic $PM_{2.5}$ due to emissions from traffic and other sources is estimated have been about 350 000 in the year 2000 (CAFÉ, 2005). These health effects are linked to high economic losses in the form of higher costs for medical treatment and losses for employers because of sickness of the workforce.

Noise — an underestimated problem

European cities have become increasingly 'noisy'; not necessarily because the noisy places have become noisier, but rather because there are fewer quiet places left. People are affected by noise from traffic, leisure activities and the general neighbourhood at all hours of the day and night. Detailed data on noise in Europe are scanty; however, a general picture is given below.

Road traffic is the dominant source of exposure in major urban areas. The *EU Thematic Strategy on the urban environment (EC, 2006d)* reports that exposure to continuous road traffic noise affected:

- 160 million people in the EU-15 (40 % of the population) at an 'averaged' level above 55 dB(A) — associated with significant annoyance;
- 80 million people (20 % of the population) were exposed to continuous road traffic noise above 65 dB(A) — associated with cardiovascular effects.

In 2002 the European Commission introduced the Environmental Noise Directive relating to the assessment and management of environmental noise. Exposure data are not currently available for all Member states. Figure 1.3 in Chapter 1 of this report presents data on traffic noise for selected cities.

Data obtained in 2008 from a questionnaire sent out by the EUROCITIES Working Group on Noise to the network's cities show that about 57 % of the inhabitants of responding European cities are living in areas with noise levels above 55 dB, and approximately 9 % experience noise levels of above 65 dB (Figure 2.16). Extrapolations of these percentages all over Europe would suggest that more than 210 million people in Europe are exposed to levels above 55 dB and 38 million to levels above 65 dB.

Due to progressive growth in traffic levels and the general urbanisation of Europe (see Section 2.3) the situation will worsen; particularly if measures at local, national and European levels are not put in place. As an example: the Randstad (area including Rotterdam, Amsterdam, The Hague and Utrecht) in the Netherlands is one of the most urbanised areas in Europe with consequent noise pollution across the whole area despite noise abatement measures previously implemented. Given this, one might assume that noise quality in other European cities is superior, which is not the case.

Data show that noise is a serious problem in Europe. Persistent high levels of noise are associated with learning difficulties, loss of memory, inability to concentrate as well as irreversible damage to health, such as heart attacks and strokes (Stansfeld *et al.*, 2005; van Kempen, 2008; Babisch, 2006; Jarup *et al.*, 2008). In the Netherlands, between 20 and 150 people every year suffer from heart attacks brought on by traffic noise (van Kempen, 2008) (see also noise impacts in Section 1.2). Gjestland (2007) reports that 'in Norway, the 'cost' of one extremely annoyed person has been estimated to be approximately EUR 1 600 per year. Due to the linearity, the 'cost' of a moderately annoyed person thus equals EUR 800 per year.'

Complex problems require smart solutions

The solution is not simple, as the situation is a complex result of our lifestyles, in particular as related to transport of people and goods. Cities and towns are lively places, where people live and work; where all kinds of economic, social and cultural events take place; where trade and industries are established and where roads converge. Although most city dwellers wish to live in a healthy environment, this is not the only demand on their living environment. Quality of life means healthy air and less noise, but also employment, possibilities for recreation such as shopping, entertainment and cultural outings. Therefore, in addition to the promotion of clean air and public health, local and regional authorities have to pursue many other objectives that contribute to the well-being of their citizens, such as economic prosperity, mobility, jobs and the preservation of the economic, social and cultural functions of inner cities. Cities cannot simply shut down all transport activities and industries in order to realise clean air and a better acoustic environment.

The challenge for cities is to find acceptable and smart solutions for environment and health problems; to strike a balance between different kinds of policies and to integrate them into a single city plan that gains public support.

Cross-border dimensions

Air pollutants, such as fine particles and ozone (precursors), can travel thousands of kilometres through the air and move from Member State to Member State and beyond; in other words, Member States export and import air pollution. This is a problem for local and regional authorities as a proportion of pollution in cities derives from neighbouring regions. In Vienna for example, only a quarter of air pollution is generated by the city;

Figure 2.16 Noise data for 52 European cities

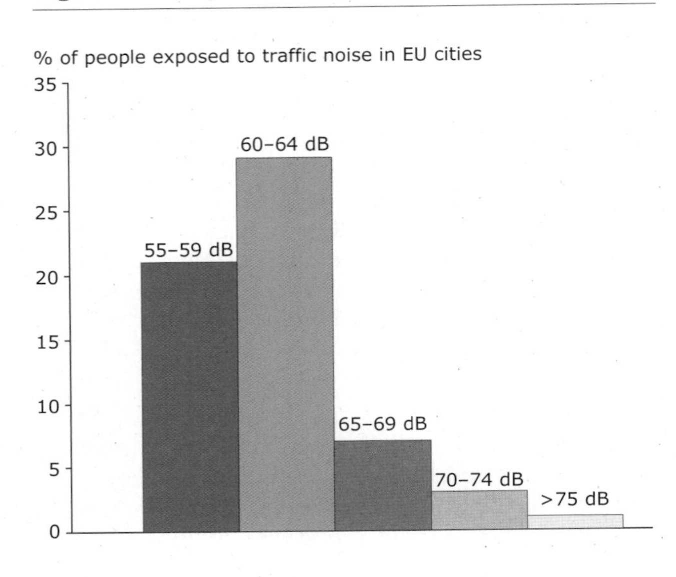

Note: The questionnaire asked how many people were exposed to what noise classes.

Source: Noise Questionnaire EUROCITIES, 2008.

the rest originates from other geographical areas. In the German city of Coburg (43 000 inhabitants), limit values are exceeded during the night when most people are asleep; this pollution is therefore not caused by urban traffic. Another challenge is the mixture of contributors to air pollution in the city, which vary from local to large-scale and can include background contributions from many other sources such as industry, agriculture, shipping, and activities in other cities, regions and countries. For particulate matters (PM_{10}) for example, up to 80 % derives from sources that are not local.

The local contributions in cities, mainly caused by traffic, form a layer over the city with highest concentrations in the streets with most traffic. Nitrogen oxides (NO_X) and particulate matter (PM_{10}, $PM_{2.5}$ and soot) emissions of cars and heavy vehicles lead to higher levels of nitrogen dioxide (NO_2) and particulate matters in ambient air. The contribution caused by traffic can be the source of half or more of the current concentration of NO_2; as in Rotterdam (Figure 2.17). Here, road traffic and shipping contribute about 70 % of NO_2, while for PM_{10}, the concentration in ambient air is mainly determined by sources from abroad. The local emission, together with the background concentrations, leads to overall ambient air concentrations above the limit values of the European Air Quality Directives (Directive 2008/50/EC) (see also Box 2.13)

Figure 2.17 Rotterdam region — contributions to NO_2 and PM_{10} concentration from different sources, 2000

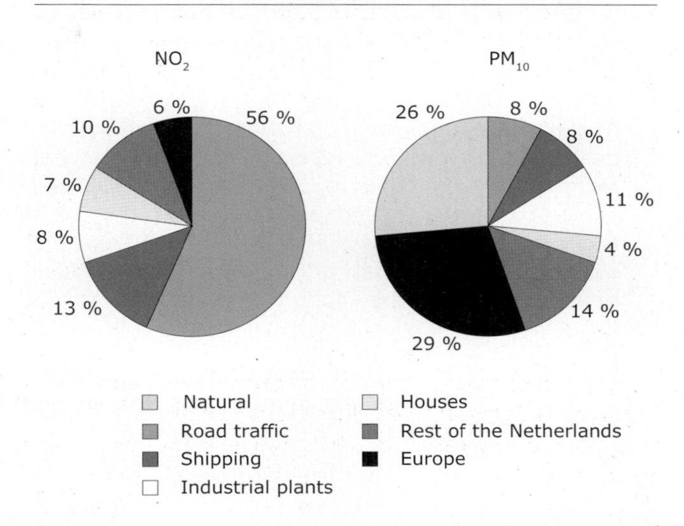

Source: The Province of South Holland, 2005.

Compared to air pollution, noise is perceived as a local and temporary problem since noise emissions mostly affect areas close to the source. Transport is the main source of noise but this derives not only from local traffic but also from regional, national and European traffic — heavy goods vehicles from the same companies can be seen everywhere in Europe; aircraft come from all over the world. Regulations of emission values are European rules, but noise has a cross-border dimension and needs to be tackled at a high administrative level.

EU policies tackling problems

As the sources of air pollution and noise and the drivers behind these sources are multiple, actions need to be taken in many sectors and at all administrative levels. In most cities, road transport is clearly the main source of air pollution and noise.

At the European level, limit values for air quality are set in the Air Quality Directives. Since the CAFÉ programme suggested emission standards for cars, ships, agricultural farms and industry, a number of actions have been taken; for example, the introduction of stricter Euro standards to reduce vehicle emissions. In addition, at the Member State level it anticipated that the revision of the National Emission Ceilings Directive will introduce stricter limits. The IPPC directive and BREF documents have also assigned limits for industrial emissions. It is important that such emission standards are adopted in time for Europe to be able meet the quality standards.

The EU Environmental Noise Directive aims to define a common approach intended to avoid, prevent or reduce the harmful effects of exposure to environmental noise. It requires Member States to determine exposures to noise in major urban agglomerations through means of noise mapping (Box 2.14). This requires Member States throughout Europe to assess the number of people disturbed during the day and at night, inform the public of the results and where necessary, prepare and adopt action plans with a view to preventing and reducing environmental noise. This information is also used to develop a long-term EU strategy to reduce the number of people affected and provide a framework for developing existing Community policy to reduce noise at source. Improved standards for vehicles, including tyres, will help with this.

Cities acting

There are numerous examples of cities and towns combating air pollution and noise. Promoting public transport, walking and cycling by calming streets and restricting road travel; and introducing parking fees or local regulation can be very effective. For

Box 2.13 Italian urban areas — air quality (8)

An analysis of urban air-quality data in major Italian urban areas with more than 150 000 inhabitants, shows that PM_{10} (particulate matter with dimension less than 10 micrometer), NO_2 (nitrogen dioxide) and O_3 (ozone) are the most critical atmospheric pollutants.

Regarding PM_{10}, exceedance of both the annual-limit value and daily-limit value, which should not to be exceeded more than 35 times in a calendar year (9), occurred in almost all urban areas, and for the majority of years between 1993 and 2006. Figure 2.18 shows the maximum number of PM_{10} daily limit value (50 µg/m³) exceedance that occurred in 24 Italian urban areas in 2006. Data show that in 13 of them more than 35 days of daily limit value exceedances were measured. Long term urban air quality data for PM_{10} show that, after a decrease in air concentrations up to the early 1990's, the effectiveness of measures and actions adopted to reduce PM_{10} pollution have only a limited effect.

Figure 2.18 Maximum number of PM_{10} daily limit value (50 µg/m³) exceedance in the 24 Italian urban areas occurred, 2006

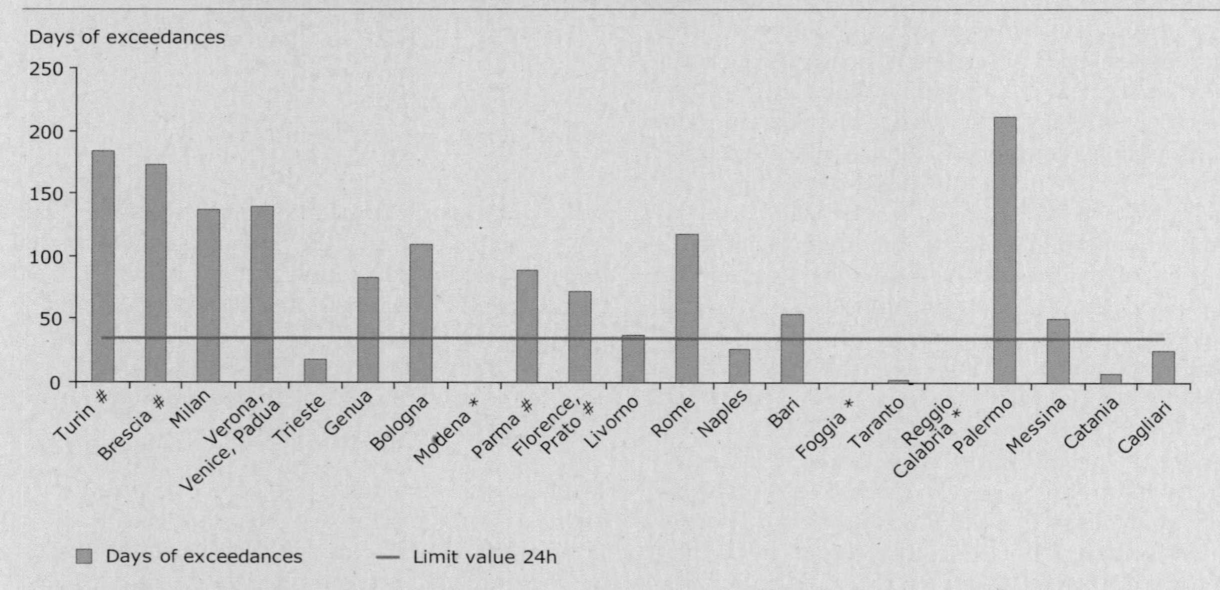

Note: # Data refer to background monitoring stations; * No data available.

Figure 2.19 shows that road transport is the biggest source of PM_{10} emissions in 19 cities (out of the 24 analysed), accounting for more than half of the emissions in 11 of them and reaching peaks of more than 60 % in cities like Rome, which scored the highest emissions levels of all.

Road transport also represents a major source of NO_x emissions in urban areas, amounting to more than 50 % of the current concentrations in 18 Italian cities. Other notable sources of this pollutant are the industrial sector (74 % and 91 % in Venezia and Taranto, respectively), domestic heating (more than 20 % in Northern cities like Milan, Brescia and Bologna) and maritime transport in seaports (i.e. 41 % in Cagliari).

Practical measures and instruments in the fields of road transport and mobility, e.g. related to car ownership or car fleet composition, can play a major role in improving the air quality and quality of life. The number of cars per 1 000 inhabitants in Italian cities, for example, shows that the values are the highest in Europe, after Luxembourg.

(8) Moricci, F.; Brini, S.; Chiesura, A.; Cirillo, M. C. (ISPRA — High Institute for Environmental Protection and Research (former APAT)).
(9) EU Directive 99/30.

Box 2.13 Italian urban areas — air quality (cont.)

Figure 2.19 PM_{10} municipal emissions per macro-sectors in 2005

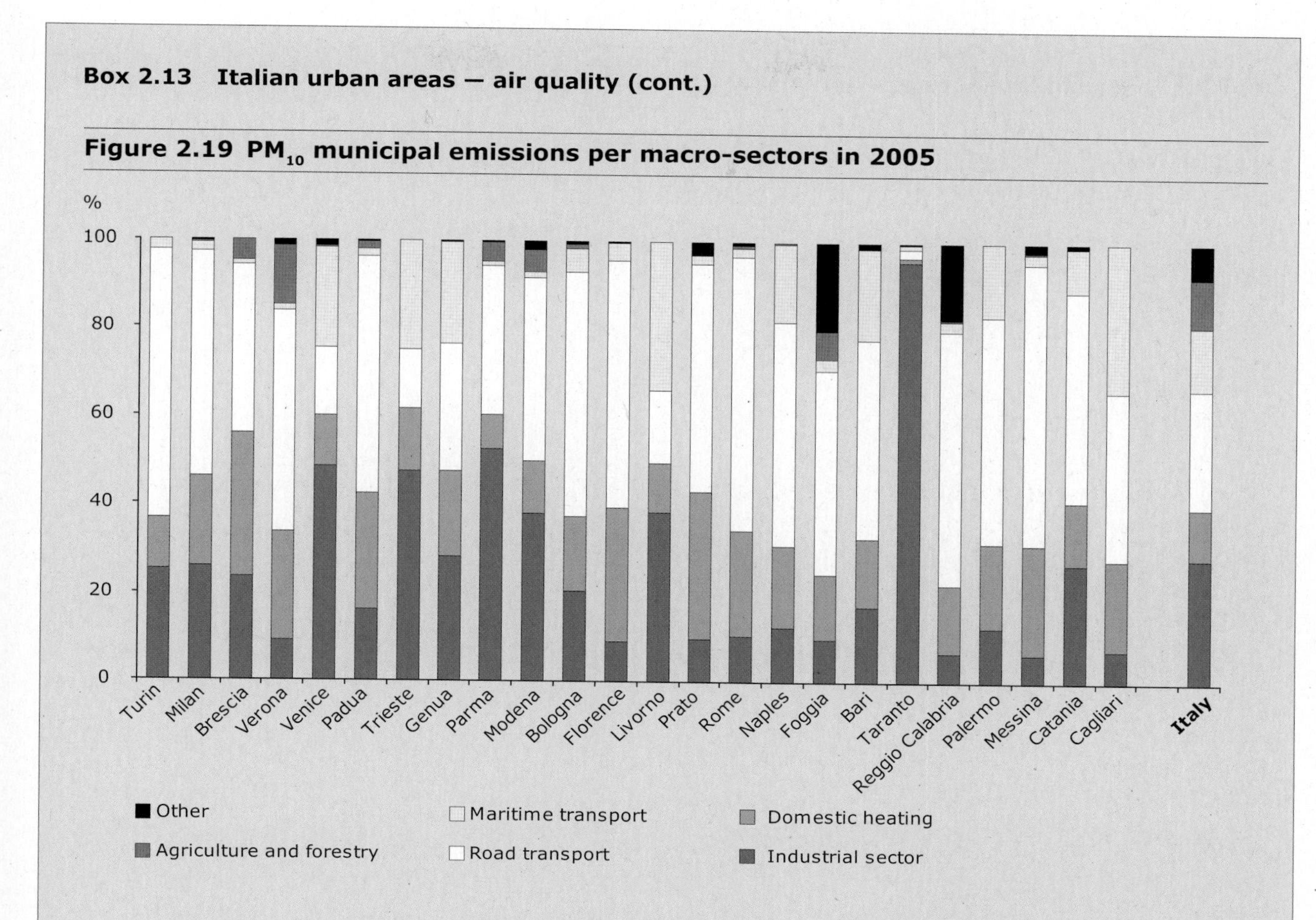

From 1996 to 2000 this value has steadily grown, and from 2000 onwards registered a slower rate of growth which has been compensated with higher mobility within large metropolitan areas. The city of Rome had the highest levels of car ownership per 1 000 inhabitants in 2000, 2005 and 2006 for the whole of Italy (Figure 2.20); the corresponding high levels of PM_{10} are clearly attributed to car usage in the city (Figure 2.19).

As far as the quality of the urban car fleet is concerned, there has been a general growth of more environmentally friendly cars responding to EU emissions-related directives, though the distribution is not homogenous across the national territory (Figure 2.21). Although this could be certainly considered a positive trend, the benefits generated might be offset by the high number of cars per 1 000 persons, the increase of diesel cars (more than 30 % of the car fleet in many cities in 2006) and of those with big engine displacement (> 2 000 cubic centimetres), and by the increase in the length of journeys due to urban sprawl.

Moreover, measures and instruments in the fields of urban transport and mobility are important but cannot solve urban air quality problems alone, as one could easily see from the non-urban contributions (for example Figure 2.19). Cities need support from European and national policy, and an integrated approach involving different sectors and different government levels.

Box 2.13 Italian urban areas — air quality (cont.)

Figure 2.20 Number of passenger cars per 1 000 inhabitants in the 24 municipalities (1996, 2000, 2005, 2006)

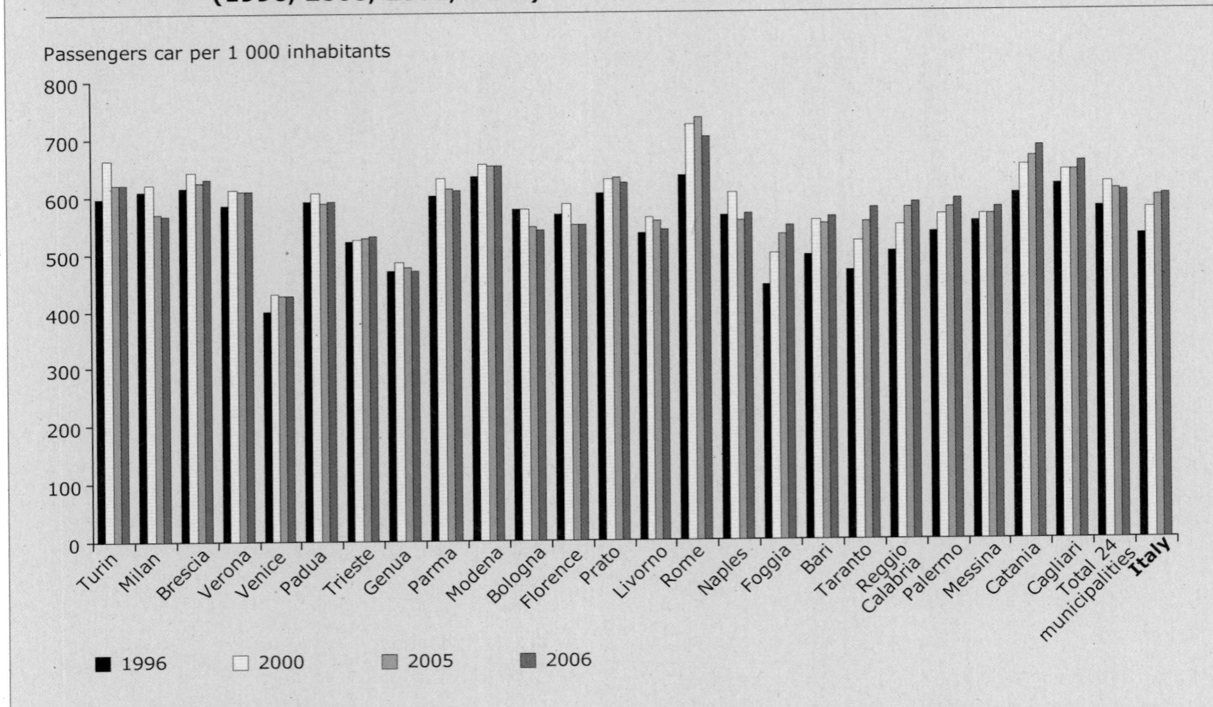

Figure 2.21 Car fleet composition per emission standards in 2006

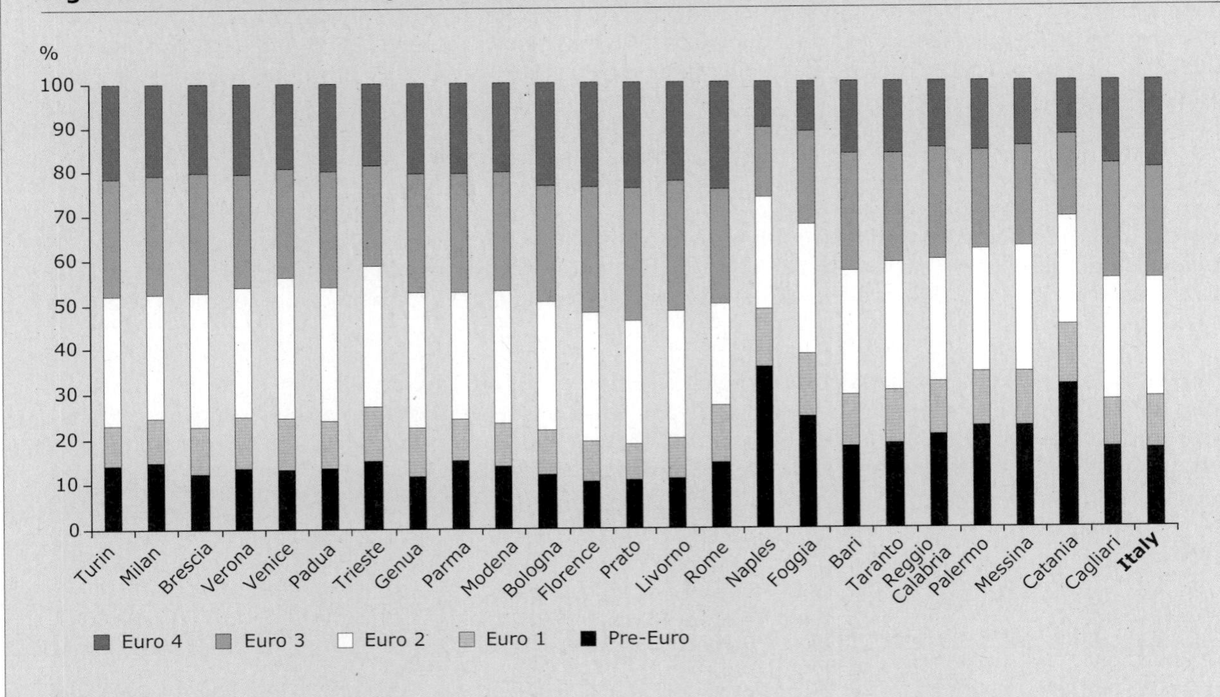

Source: APAT, 2007.

example, in London the congestion charge greatly reduced the number of cars in the city. However, air quality was not improved measurably for some substances (PM and NO_2). Therefore, in 2008 London introduced a low emission zone for a much larger area, in which only lorries with a well-defined emission standard are permitted. In Copenhagen the promotion of a walking and cycling infrastructure and very restrictive parking policies in the centre has led to 36 % of Copenhageners now choosing to cycle to work (see also Box 2.6). Local administrations plan compact cities enabling the reduction of transport demand and shifts to public transport, walking and cycling. In such cities, the objective is to ensure that everyone is within 1 km of green space — 5 minutes by bike or 10 minutes on foot. A good example is Stockholm (Map 2.9) where the 'red fingers' of the town sit side by side with the 'green fingers' — rivers, parks and other green spaces.

Over the longer term inner cities can be built more compactly, based on closed blocks rather than single buildings, so helping protect open spaces against noise from sources at the surface (Box 2.14). Furthermore, technical measures like noise barriers or tunnels support the reduction of noise and at the same time help limit local air pollution. These solutions can be developed to meet specific local problems of air quality and noise. However, the emissions in total remain the same, as they do not address the problem at source. Also, such measures often incur high and permanent maintenance costs (Box 2.14).

To further improve air quality, some cities invest in improved insulation of houses and efficient, low emission heating systems like district heating (see example in Box 2.4). Energy management of houses and combustion plants is a means to secure low emissions in buildings. Green public and private procurement and the procurement of clean vehicles, environmentally certified buildings and applications offer local government opportunities to demonstrate good practice to citizens.

Map 2.9 Stockholm, Sweden: green and red finger zoning plans

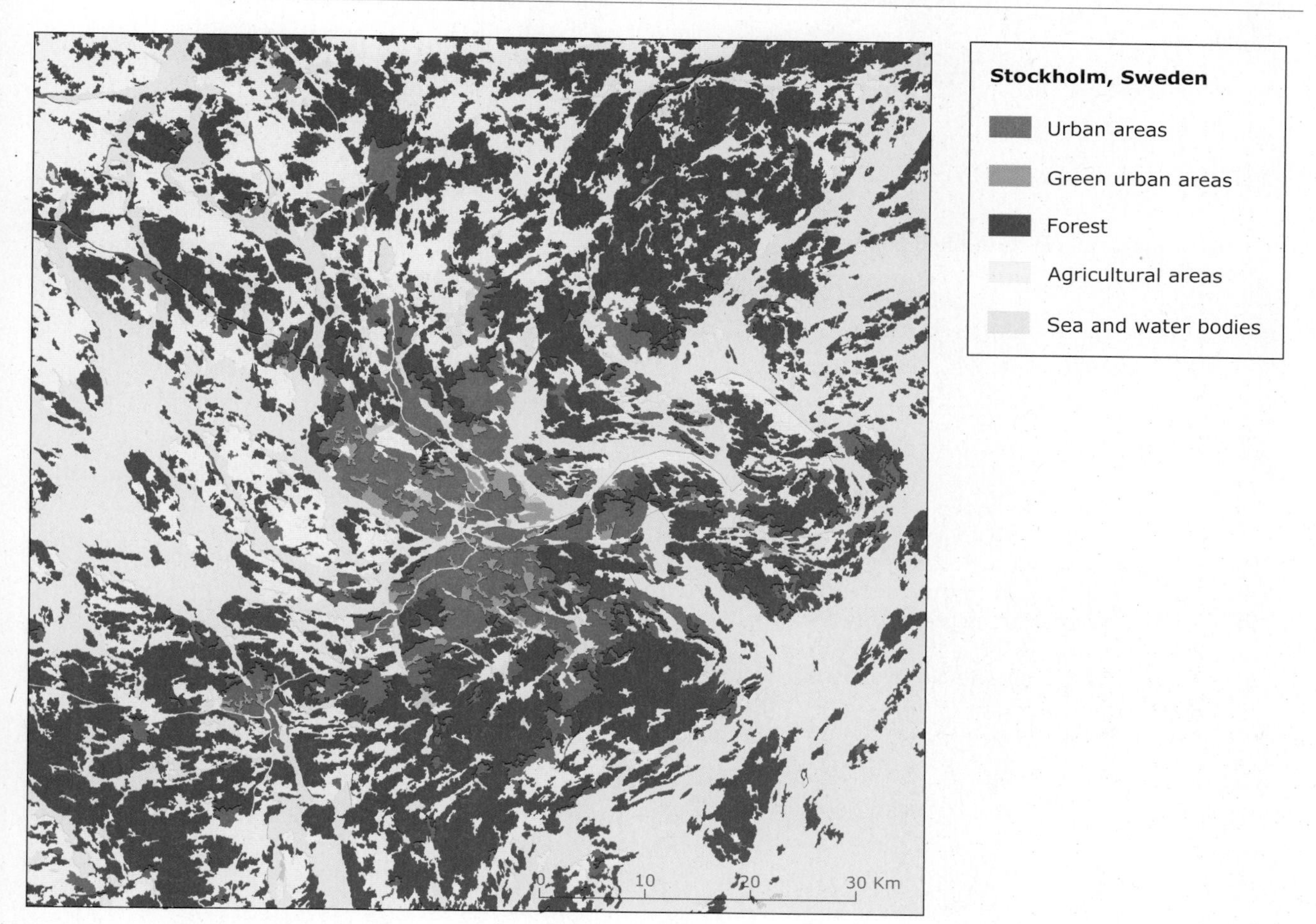

Source: EEA, Corine Land Cover 2000.

Box 2.14 Rotterdam — examples of protection against noise and air pollution

Map 2.10 Traffic noise map of Rotterdam, the Netherlands

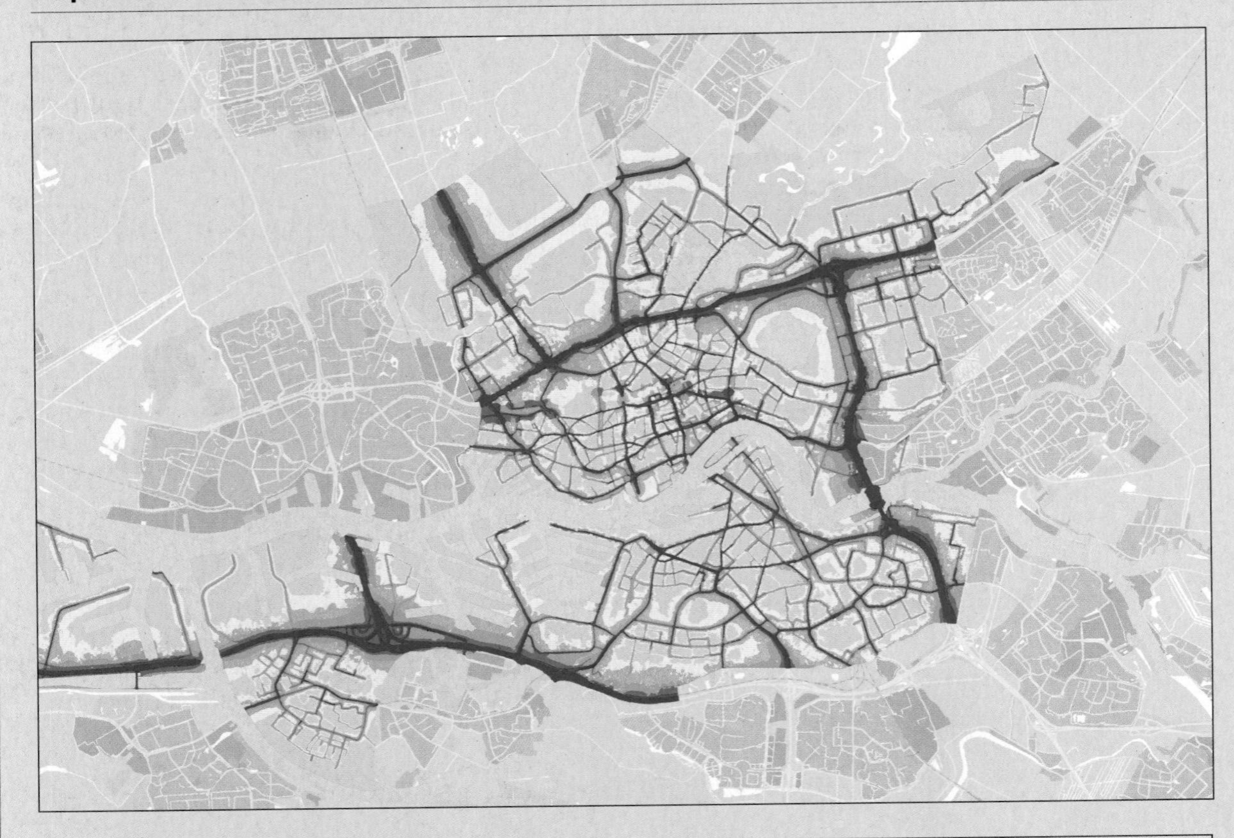

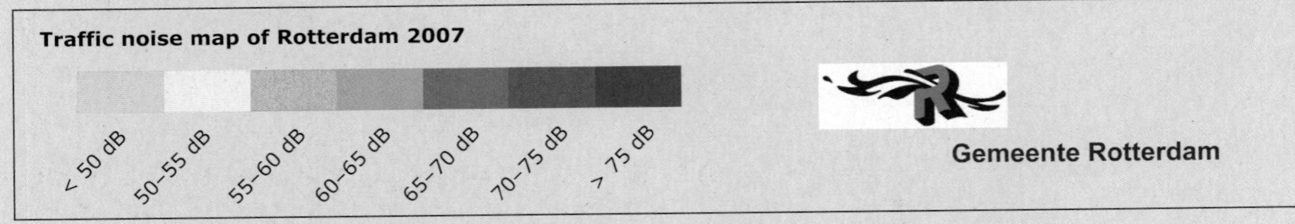

Source: DCMR, 2007.

The Map of Rotterdam shows very high and high noise level on nearly every major road.

Map 2.11 shows a typical situation where there is a high level of noise at the front of the dwellings, but the backs are quiet due to the compact building frontages. The noise action plan for Rotterdam proposes the use of quiet road surfaces in this area, and future replacement of the road with a tunnel.

Such a tunnel will also affect the air quality situation as described in Maps 2.12 and 2.13.

Box 2.14 Rotterdam — examples of protection against noise and air pollution (cont.)

Map 2.11 Closed blocks keep noise out

Source: DCMR.

Maps 2.12 and 2.13 demonstrate the improvement of air quality when tunnels are used for large roads in residential areas. However, there is a problem at the entrance and exit of the tunnels, where the air quality can be worse than without the tunnel (left figure). At this point, there either needs to be a larger area where no one is exposed to the polluted air or the adoption of an expensive technical solution, such as cleaning the air flow by filters. The maps show the results of a calculation on NO_2 concentrations made by TNO.

Map 2.12 No tunnel

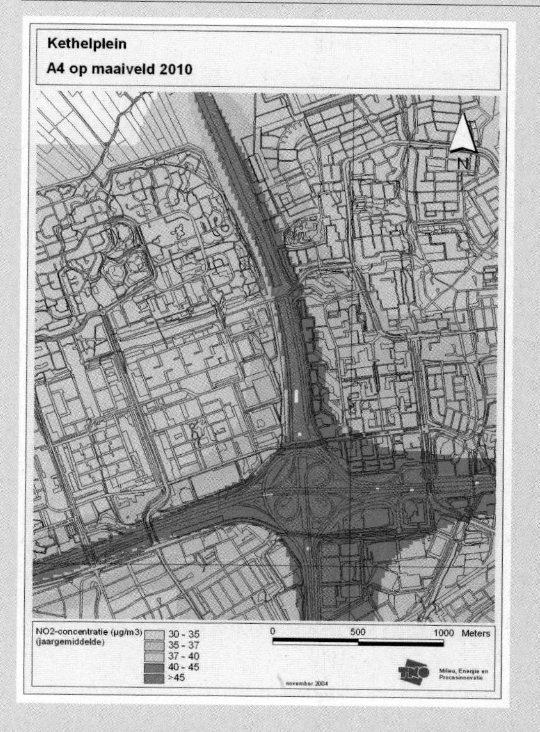

Source: TNO, 2003.

Map 2.13 With tunnel

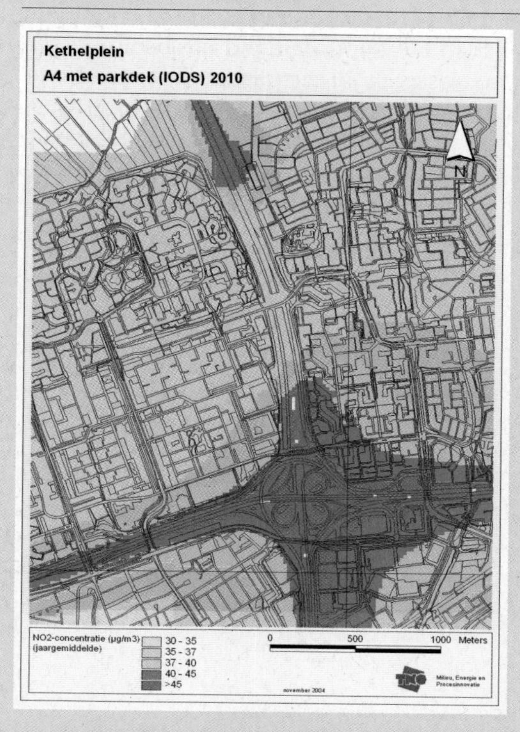

Source: TNO, 2003.

Barriers to effective action

Despite all that is known and has been done, air pollution and noise in cities remain serious problems. This is because while air pollution and noise are recognised as major public concerns, changes in the organisation and structure of urban areas to reduce air pollution and noise are not always popular. Individual interests collide with societal interests to provide clean air and quiet environments for all citizens. For example, individual car owners want to use their car, but better air and acoustic quality for all citizens requires more environmentally friendly modes; or, business urges policy-makers to provide more road access presenting it as a basic condition for investment in jobs. These conflicts are difficult to resolve at a local level.

The case study of Italian cities (Box 2.13) and the case of the German city of Coburg mentioned earlier demonstrate clearly that the problems of air pollution cannot be solved by one administrative level alone. Too many sources of air pollution lead to a high background concentration, which often requires relatively small contributions from local roads before limit values are reached. A similar situation applies to noise, in particular from transit transport. The improvement of air quality must be a joint effort by all government levels; otherwise cities will not be able to meet the standards for air quality and ambient noise. While road transport demand has been increasing progressively in the past decades — in particular in the new EU Member States — in many cities the speed of introduction of responses to new Euro standards for vehicle emissions was too slow to meet the air quality limit values of the EU directive before they became even stricter.

Overcoming barriers to action

Local, regional, national and European policies must go hand in hand to achieve the values of the Air Quality and the Ambient Noise Directives. Only an integrated approach will be successful. This means integration of policy, legislation and measures at all levels (local, regional, national and European) and extended beyond air and sound quality to include energy, safety, urban design, public space etc. It is also crucial to include business and industry in this approach.

The European level

Europe as a whole must set the framework, provide basic conditions and aim to reduce background emissions. This must be done primarily by setting emission standards for vehicles, including ships and aircraft, and industry; exposing the external costs of noise and pollution sources; and providing incentives for cleaner and quieter alternatives, including the shift to other transport modes, and support for the introduction of environmental zones and other local environmental policies.

European legislation should also enable practical and flexible solutions. In the case of tunnels (Box 2.14), air quality can be improved significantly where roads are driven underground but levels are typically exceeded elsewhere. EU legislation does not permit such measures as the air pollution at the end of the tunnel increases above the limit value. Local and regional authorities should be encouraged to adapt European standards to the specifics of the local geographical, environmental, social and economic situation. Only then, they will they be able to tackle air pollution effectively.

National policy

Policies at national level can also influence individual choice, for example whether to drive a car at all and if so what type of car. Measures to influence these choices could be based on taxation according to the pollution levels of the vehicle, and the provision of alternatives to the car such as public transport.

Cities

At a local level cities need to strengthen their efforts to make the available good practice to reduce air pollution and noise levels not only for their cities, but also to support mainstreaming throughout Europe. This requires working with other sectors — otherwise it remains an isolated solution that cannot be fully developed and will not achieve its full potential. Major areas of necessary cooperation include urban planning to reduce sprawl, decrease transport demand, and facilitate the construction of more compact cities (see also Section 2.3). Public participation is also essential to ensure the representation of the interests of all societal groups. In these ways air and sound quality policies can become fully integrated into urban and transport planning.

Strong political commitment

The improvement of air and acoustic quality in Europe's cities requires a strong political commitment and a shared future vision for the city and all citizens. The vision facilitates the assessment of single measures and their effects in a wider context and provides the platform to convince people of the benefits of changes that are initially unwelcome. Achieving good air quality and acceptable noise levels while maintaining socio-economic and cultural infrastructure requires a well-balanced approach and cooperation at all administrative levels and of all stakeholders.

2.5 Climate change

In 2008 Barcelona ordered huge quantities of water delivered by tanker to serve its population and tourists. In 2003, the summer heat wave killed 14 800 people in France, 18 000 in Italy, and all together around 52 000 across Europe (EPI, 2006). In 2002 pictures of flooded Dresden and other German cities showed extreme flooding of the River Elbe. All such events may occur more frequently because of climate change.

The impacts of climate change are increasingly a serious threat to people's quality of life; but our lifestyles, the way we consume and produce goods and services, continue to trigger further climate change. As a key policy objective, the EU has stated that to avoid major irreversible impacts on society and ecosystems the temperature must be stabilised to below 2 °C above pre-industrial levels. Global action to reduce greenhouse gas emissions is needed, but even if the EU target is achieve, impacts of climate change will persist and we will need to adapt to these new conditions.

The concentration of population and activities in urban areas means that cities and towns must play a major part in mitigating climate change both locally and globally. Europe requires that cities contribute to the battle against climate change but also needs to complement and support actions the cities take. This is a twofold challenge for cities as they will also have to adapt to the effects of inevitable climate change.

Cities and upcoming changes

In addition to a rise of the mean annual temperature in Europe (Map 2.14), projections indicate an increase in the severity and frequency of droughts, floods, heat waves, and other extreme weather events that are expected to have major impacts during this century (IPPC, 2007; EEA, 2008). Also, as Map 2.15 shows, the expected impacts differ widely across the European regions.

In coastal areas sea levels are predicted to rise between 10 and 45 cm and by 2050 many cities face the serious risk of flooding. Some countries, for example the Netherlands, may have the knowledge and resources to protect their coastlines, but others may not be so fortunate and will require ongoing support and guidance. Even for those cities that have some knowledge or experience in flood risk management, the potential severity of some predicted impacts means that without innovative solutions, the effects may be unmanageable. Map 2.16 shows the risks of flooding in other areas of Europe due to extreme weather events.

Droughts and heat waves are most associated with the southern parts of Europe. However, a simple geographic division of threat will not suffice explanation, as shown by the Paris heat waves of recent years. The effects of climate change are also dependent on the specific characteristics of the locality. For example, the 'urban heat island' effect is a well-known result of urbanisation. The case of Zaragoza (Spain) shows that differences in urban density and vegetation cover account for 37 % of the thermal variation between the city and its surrounding rural areas. Temperature also differs across the city, with green urban areas clearly cooler than high density urban areas (Cuadrat Prats *et al.*, 2005). Another example is vulnerability to floods: weather events play a part but so does the way urban areas are built, as shown in flood simulations for the Dresden-Prague corridor (Box 2.8) and Stockholm and Göteborg (Box 2.15).

The new risks of droughts, heat waves and floods further exacerbate the existing environmental problems of many cities and towns, including low air quality and water supply problems. High population density and their physical structure make cities highly vulnerable to the impacts of climate change.

Climate change and quality of life

Climate change will have significant impacts on the environment, public health and the economy. Climate change will cause deaths during heat waves, increase health problems as a result of additional particle emissions during droughts, exacerbate ozone and air quality related health problems, and intensify the distribution and spread of infectious diseases. It will also affect the basic elements of life and hence our economy. The Stern Review argues that if no action is taken, the overall costs and risks of climate change will be equivalent to losing at least 5 % of global GDP each year worldwide. In contrast, the costs of action — reducing greenhouse gas emissions to avoid the worst impacts of climate change — can be limited to around 1 % of global GDP each year (Stern, 2006).

Map 2.14 Apparent southward shift of European cities — due to climate change, 2070–2100

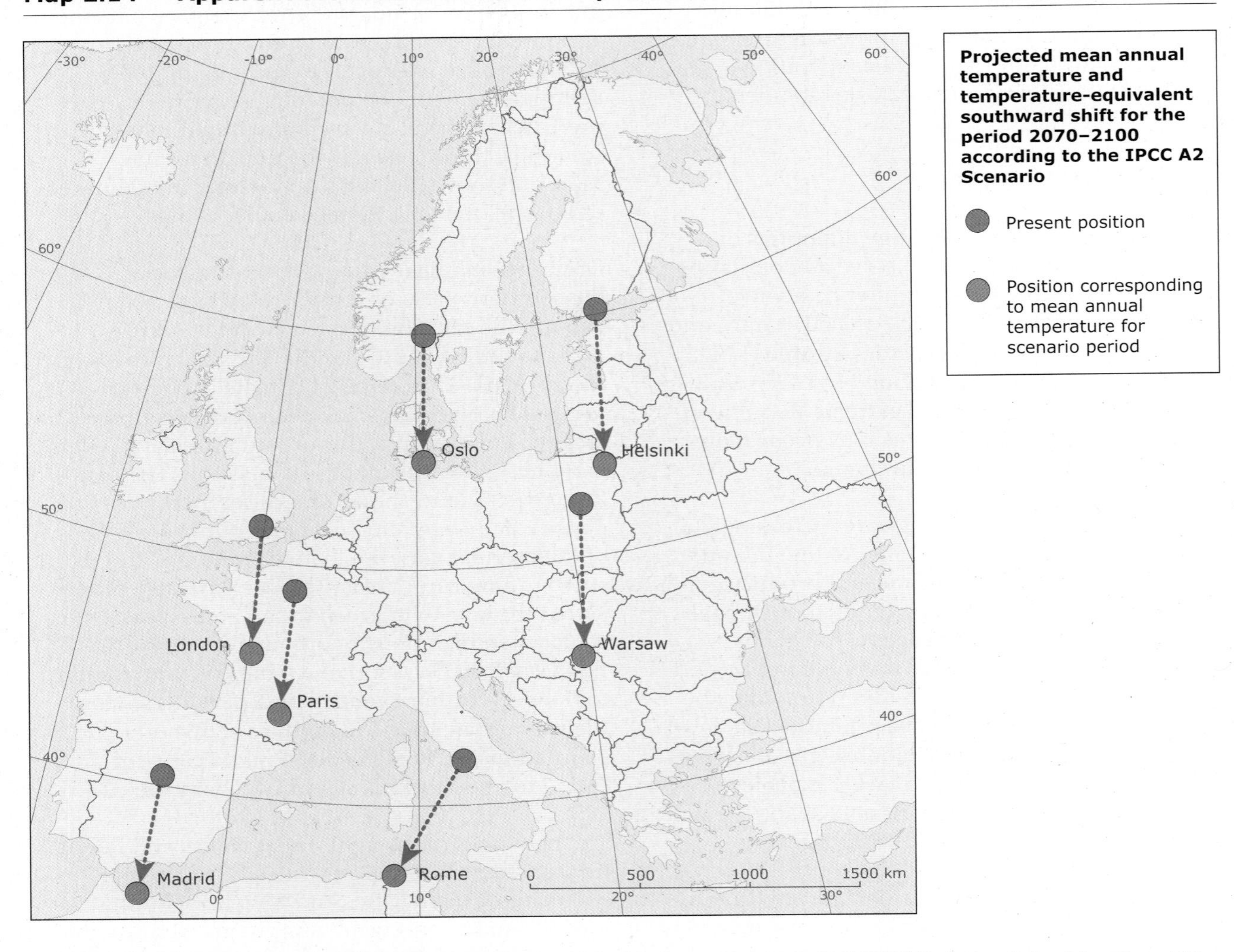

Source: Hiederer *et al.*, 2009a — Original data source: Danish Meteorological Institute, PRUDENCE Project — Data elaboration: JRC.

The burdens and benefits of climate change are not equally distributed (EEA, 2008). The type and location of threat will have significant economic implications, as some cities will suffer great impacts and others will enjoy positive effects. Within urban areas climate change can aggravate social inequalities, as typically the poor live in climatically less favoured areas and do not have the resources to adapt their housing to deal with the effects of climate change. The consumption patterns and lifestyles related to our quality of life also drive climate change and threaten the ecological, economic and social basis of quality of life in the longer term.

However, apart from its serious negative effects, climate change and the fight against it can create new opportunities to develop the local and regional economy and new employment through, for example, the development of the global market for new technologies in efficient energy production, renewable fuels and heating. The vulnerability of cities and increasing awareness are a driving force to find innovative solutions for adaptation to climate change and ensure quality of life. Financial benefits from a shift towards the development of new technologies could, at least in part, balance the costs of necessary *changes in production and consumption.*

The role of cities in the mitigation of change

Emissions of greenhouse gases are linked to the material consumption of goods and services, and in particular the fossil energy resources used to produce these goods and services and make them available to the consumer (see Section 2.2). The urban population in Europe accounts for 69 % of European energy use (IEA, 2008) and thus most greenhouse gas emissions;

Map 2.15 Climate change impacts for the main biogeographic regions of Europe

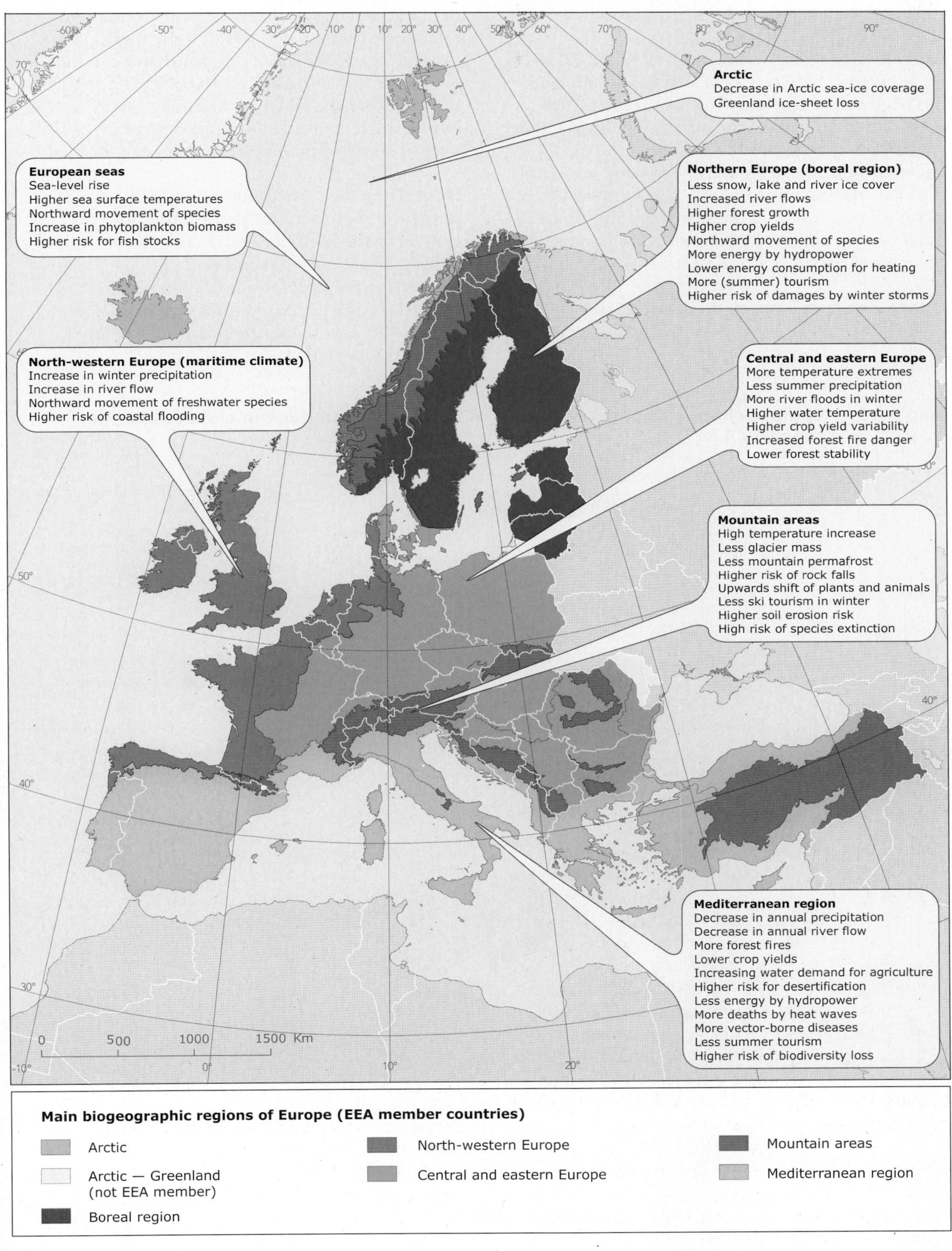

Source: IPPC, 2007; EEA, 2008.

although on average, because of the effects of population density, the individual city resident consumes less energy than the rural resident.

Overall, final energy consumption in EU-27 has risen by 10 % between 1990 and 2006. Transport has been the fastest-growing sector since 1990 and is now the largest consumer of energy. Urban transport alone accounts for 40 % of the CO_2 emissions produced by European road transport (EC, 2007d). EU-wide energy projections anticipate a continued growth in energy consumption to 2030 in all sectors and hence increasing emissions of greenhouse gases (Figure 2.22).

Underlying the challenge to maintain quality of life in urban areas is the need for sufficient and sustainable supplies of energy to provide the economic activity underpinning increasing energy demand and expectations. This is becoming more and more difficult in times of tight energy markets, expanding global energy demand and complex geopolitical circumstances. In order to mitigate the various risks while maintaining quality of life, efforts must continue to reduce the urban as well as overall demand for energy and energy services.

Despite the fact that today, most of the emission reduction plans and measures are under the control of the Member States, and implementation takes place at national and regional level, cities have by virtue of their population size a great potential and specific competence in mitigation policies; in particular, the potential to plan the

Map 2.16 Climate change impacts — exposure to flood risk under the climate change scenario A2

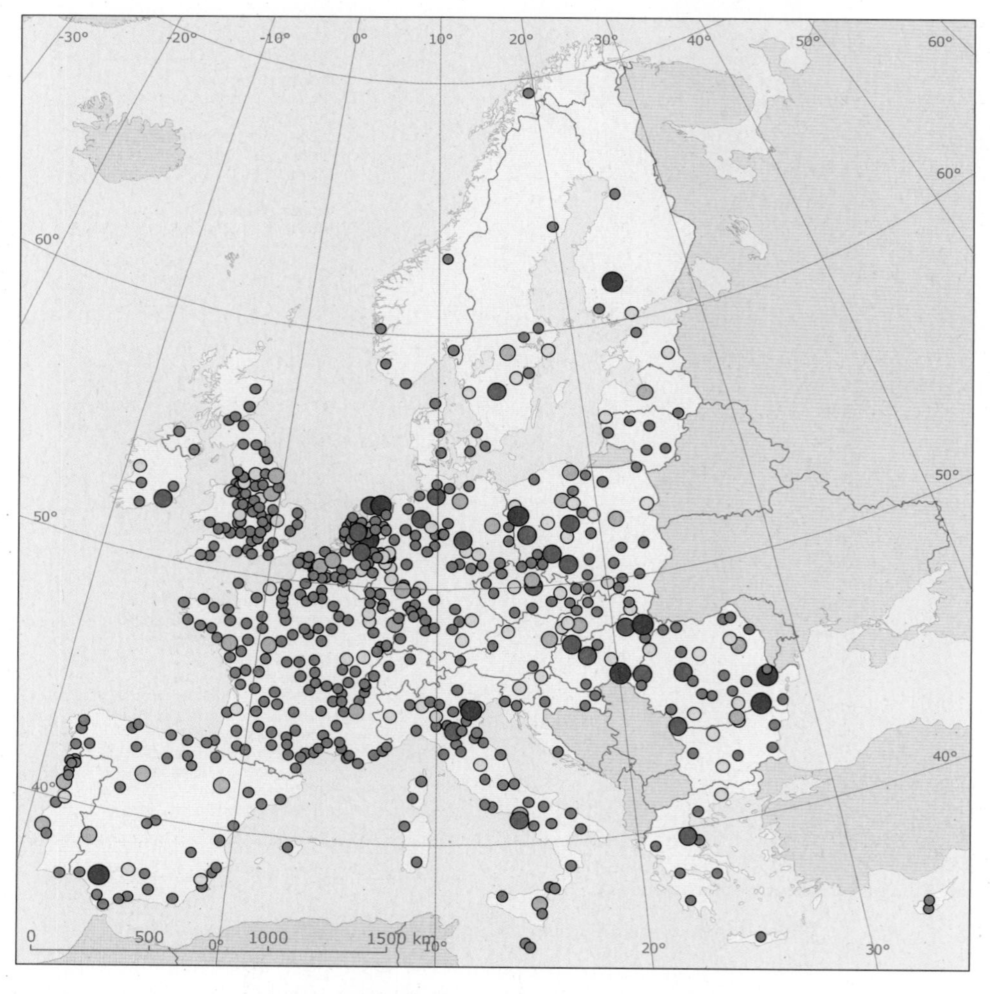

Note: Several major European cities (> 100 000 inhabitants) are potentially exposed to flood events (return period 100 years). This is a no-action scenario and coastal floods are not included.

Source: Hiederer *et al.*, 2009b.

Box 2.15 Threat of flooding — examples from Stockholm and Göteborg (Sweden)

Lake Mälaren runs into the sea at Stockholm and the Vänern Lake at Göteborg. They provide drinking water for these cities. The consequences of flooding corresponding to the 100-year return level would be:

Mälaren:

- Buildings would be flooded (housing, offices and service = area of 360 000 m^2, other buildings = area of 480 000 m^2)
- Tunnels might be flooded, e.g. the Riddarholms-tunnel through which all train traffic through Stockholm passes, tunnels for water, electricity and telephone.
- Contaminated areas would be flooded leading to leakage of harmful substances which might affect drinking water quality.
- Roads and rail roads would be flooded.

Photo: © Jens Georgi

Vänern:

- Buildings (housing, offices and service = area of 1 200 000 m^2, other buildings = area of 1 500 000 m^2), roads and rail roads would be flooded.
- Shipping might be cancelled between the lake and the sea.
- Contaminated areas would be flooded leading to leakage of harmful substances which might affect drinking water quality.

Source: Miljödepartementet, 2006.

Figure 2.22 Final energy consumption by sector in EU-27

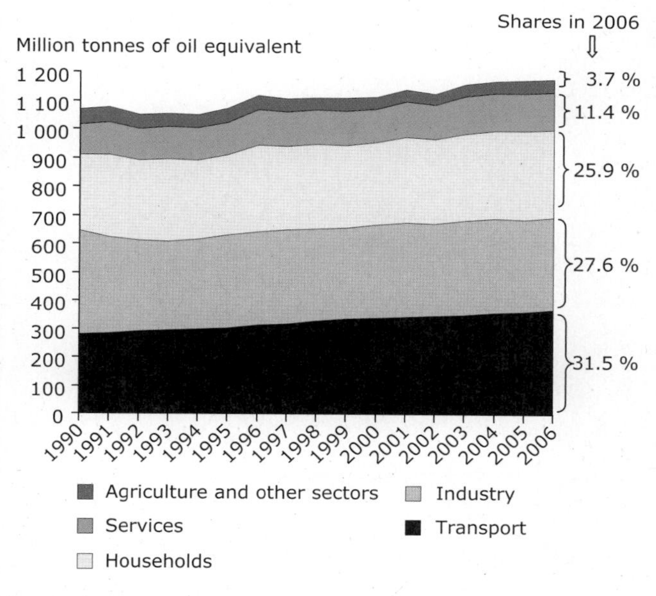

Source: EEA; Eurostat, 2008.

city in a way that facilitates sustainable urban transport, low energy housing, etc. City design should facilitate further lowering of average individual energy consumption. New technologies for energy efficiency and renewable resources, such as solar energy, wind energy and alternative fuels, are also important, as is the provision of opportunities for individuals and organisations to change their behaviour. Frontrunner cities are already beginning to act as catalysts for change and provide excellent examples of best practice (see Boxes 2.16 and 2.17).

Reducing energy consumption

In response to the ratification of the Kyoto Protocol and also to increasing concerns about the security of energy supply, European Member States have adopted various national programmes aimed at reducing greenhouse gas emissions, and various policies and measures have been adopted at the EU level, in particular through the European Climate Change Programme, for example:

Box 2.16 Barcelona (Spain) — Plan for Energy Improvement(PEIB)

Initial situation

In Barcelona it has been recognised that a shift towards sustainable energy systems in cities is urgently required. The promotion of a rational use of energy, together with the development of renewable energy strategies became a clear priority in Barcelona.

Photo: © Agència d'Energia de Barcelona

Solution

The City Council established the Plan for Energy Improvement in Barcelona (PEIB) covering the period 2002–2010 with the following goals: to increase the use of renewable energy (especially solar energy); to reduce the use of non-renewable energy sources and to lower the emissions produced by energy consumption. The plan comprises promotion policies, demonstration projects, legal and management instruments, and the integration of energy measures into urban development.

A relevant initiative within the plan has been the further implementation of a Solar Thermal Ordinance, which was approved previously and aims at regulating, through local legislation, the implementation of low-temperature systems for collecting and using active solar energy for the production of hot water for buildings. New buildings and buildings undergoing major refurbishment are required to use solar energy to supply 60 % of their running hot water requirements. Since its enforcement until the end of 2006 a total of 40 095 m^2 of solar panels have been installed with annual savings of 32 076 Megawatt hours per year, corresponding to the amount of energy needed to provide hot water for 58 000 inhabitants per year.

To promote the Ordinance and its acceptance, Barcelona has implemented a broad communications program. The city has published an explanatory guide for the Ordinance in several languages, held periodic round tables and meetings with stakeholders (contractors' associations, engineers, architects, environmental organisations, neighbourhood associations, citizens), promoted the Ordinance in neighbouring cities, implemented demonstration projects (such as solar thermal installations at swimming pools), and supported community based initiatives such as the 'Solar Day' in Barcelona.

Results, lessons learnt and transfer potential

As a result of the PEIB the generation of renewable energy produced by both thermal and photovoltaic solar power installations in Barcelona increased dramatically. The communications program is crucial to encourage adoption of the different measures and of the Solar Thermal Ordinance. The plan is turning Barcelona into one of the cities that make most use of solar energy and its Solar Thermal Ordinance has become a model for more than 50 Spanish municipalities and was a major input to the new Spanish building code.

Despite these remarkable achievements, further efforts are necessary to reduce and reverse the tendency to increase energy consumption.

Figure 2.23 Renewable energy, Barcelona (Spain)

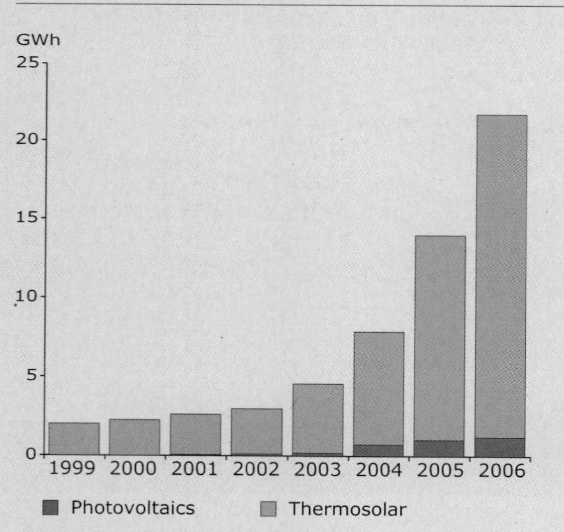

Source: Agència d'Energia de Barcelona.

Source: Agència d'Energia de Barcelona, http://www.barcelonaenergia.cat; http://www.bcn.cat/mediambient.

Box 2.17 Västra Hamnen in Malmö (Sweden) — carbon neutral residential area

Initial situation
Structural changes in Malmö's economy have transformed the city from its traditional industrial background. The city decided to build a new attractive and sustainable residential area in a former harbour area. Part of the concept was the aim to provide 100 % of the energy for the area from locally renewable sources.

Photo: © Birgit Georgi

Solution
In the project area 1 000 homes get their energy supply from renewable sources; solar energy, wind power and water, the latter through a heat pump that extracts heat from seawater and an aquifer — a natural water reserve in the bedrock that facilitates seasonal storage of both heat and cold water. 1 400 m^2 of solar collectors, placed on top of ten of the buildings complement the heat produced by the heat pump to supply the area. A large wind power station (2MW) placed in Norra Hamnen (the north harbour) and 120 m^2 of solar cells produce electricity for the apartments, the heat pump, fans and other pumps within the area.

The 100 % renewable energy equation is based on an annual cycle, meaning that at certain periods of the year the city district borrows from the city systems and at other times the Västra Hamnen area supplies the energy systems with its surplus.

An important part of the concept is low energy use in the buildings. Each unit is only allowed to use 105 kWh/m^2/year, including household electricity. There are many technical solutions to achieve this.

Urban density and a sustainable transport concept complement the activities to contribute to the mitigation of climate change

Results
The project has been a major success. Today, after a few of adjustments, most houses have reached the target or are close to it. The energy system has worked excellently from an overall perspective and the new technologies, where solar collectors are connected to the district heating system grid, have worked well. The only problem arose in the areas with heat pumps, a generally known technology. Seven years after its inauguration the area still attracts thousands of international visitors underlining its international significance.

An important factor in the success of the project was early and open dialogue with the construction companies. Together common goals were established with which everyone could agree, instead of relying on legislation.

More information: www.ekostaden.com.

- the EU Greenhouse Gas Emission Trading Scheme, which forms the cornerstone of EU efforts to reduce emissions of large industries cost effectively;
- increased use of renewable energy sources such as wind, solar or biomass (Directive 2001/77/EC) and combined heat and power installations (Directive 2004/8/EC);
- improvements in energy efficiency in, for example, buildings (Directive 2002/91/EC), industry (Directive 2006/32/EC), household appliances (Directive 2005/32/EC).

More information can be obtained from http://ec.europa.eu/environment/climat/climate_action.htm.

In March 2007, the EU leaders endorsed an ambitious climate change and energy plan to reduce EU greenhouse gas emissions by 20 % by 2020 as compared with 1990 levels.

Low carbon options for cities include planning efficient city structures, controlling urban sprawl, developing efficient public transport, and increasing the production and use of renewable energy (Boxes 2.16 and 2.17). It is also essential that local and regional governments adopt more ambitious local and regional targets to bring down CO_2 levels. Some cities for example Rotterdam, the Hague, London (Box 3.6) and Newcastle have made commitments to become carbon neutral. City administrations working with sectoral partner organisations are promoting reduced energy use, renewable zero emission energy and energy efficiency to mitigate the negative impacts of climate change.

Acknowledging the fact that a city alone cannot tackle the challenge of climate change, cities are also developing joint actions to mitigate climate change, for example the Nottingham Declaration on Climate Change (Box 2.18) and the Covenant of the Mayors Initiative (Box 1.8).

Adaptation to climate change

The key question in adaptation to climate change is how cities and regions can secure the functioning

Box 2.18 Nottingham Declaration on Climate Change

Initial situation
Climate change poses a genuine threat to our planet. The scale of the challenge means that all sectors of the community have to be involved if we are to meet targets for reducing emissions and adapting to climate change. Local Authorities in particular have a crucial role to play in responding to this challenge.

Solution
The declaration is a voluntary pledge to address the issues of climate change. It represents a high-level, broad statement of commitment that any council can make to its own community. The declaration was originally launched in October 2000 at a conference in Nottingham with 200 leaders, chief executives and senior managers of UK local government. To mark the fifth anniversary of the declaration it was re-launched on 5 December 2005 at the Second National Councils Climate Conference. The new version of the declaration is broadly similar to the original, but better reflects current thinking. The process of revising and re-launching the declaration was undertaken by a steering group that includes all the main national agencies concerned with the different aspects of climate change along with IDeA, LGA, Nottingham City Council and ICLEI, as well as the worldwide association of local governments concerned with sustainability.

Results
So far, over 200 local authorities have signed the declaration but it is vital that all local authorities commit themselves to the process. The declaration is an important starting point, but local authorities are encouraged to develop an action plan to ensure that their good intentions turn into reality. This new declaration is accompanied by an online action pack that outlines the milestone activities that should be undertaken, together with a range of options on how to do this.

More information: http://www.energysavingtrust.org.uk/nottingham.

of essential infrastructure for energy provision, electricity and heating, wastewater and water distribution, reduce of health risks, and avoid loss of biodiversity, green open spaces and space for the production of food, while the risks presented by storms or floods increase. The upcoming unavoidable climate changes therefore require cross-sectoral thinking and new strategies. Public spaces including squares and parks may need to be used differently and will require mechanisms for cooling and ventilation (Box 2.19). Buildings must be able to cool and heat more efficiently. Space for locally produced energy sources must be found. Cities vulnerable to drought or excessive rainfall need to act in tandem with their regions to increase water storage capacity. Others may need to adapt their city structure to rising sea levels (Box 2.20). Cities do acknowledge the need to rethink the nature of the urban fabric but concrete action is still very limited.

At the European level, the Green Paper on adaptation to climate change (EC, 2007f) and a White Paper (EC, 2009) as well as various plans at national level provide general adaptation strategies and options.

Barriers to effective action

Current efforts may be at a scale previously unimagined but are still not enough to respond adequately to the problems of climate change.

- Individual cities may feel that acting alone will have limited effect and that they are not in a position to implement all the changes they need to make. Climate change is a global challenge and must be addressed through the United Nations, the EU and national governments, so the actions taken by cities also depend on national actions and actions by other cities and regions worldwide.

- The role of urban areas in mitigating climate change is not yet reflected in current policy documents, such as those within the framework of the European Climate Change Programme, and is rarely included in national strategies and plans related to the Climate Change Framework Convention and the Kyoto Protocol.

- Coordination of many different sectors and actors poses another challenge. There are a number of existing technical solutions to reduce greenhouse gas emissions, for example enhanced renewable energy and energy efficiency, but there is no single major solution. Multiple coordinated actions require a broad integrated approach.

- The lack of understanding and knowledge present other challenges. In general, there is only a low level of awareness in cities of the concrete impacts of climate change and what they must do to reduce emissions and adapt to the impacts. The Eurobarometer (2008b) showed that 75 % of the population believes that climate change is a very serious problem, but only just over half of the people feel informed about the causes and consequences. In addition, precise information on energy consumption, greenhouse gas emissions and their sources exist for only a few cities, and this clearly complicates target setting and action.

- Current policy lacks sufficient level of long term commitment, in particular economic, to initiate the necessary measures, although awareness of the likely effects of climate change has grown since the publication of the IPCC reports (IPCC, 2007) and the Stern Report (Stern, 2006).

- Only a few of the most advanced cities have experience of implementing measures that allow adaptation to the 'new' climate change conditions (Box 2.20). The development of solutions and the exchange of experience and best practice in these areas are most valuable, but are often limited to small scale pilot projects.

Overcoming barriers by cooperation

Europe cannot expect to achieve its major climate change objectives without contributions from the major European urban centres towards achieving these goals, said Ronan Dantec, Vice Mayor of Nantes, in 2008.

As tackling climate change requires many different actions across many sectors, involving different actors at all levels, connections with policies other than climate change are critical. An integrated approach is required that links policies on air quality, road safety, noise, energy, urban sprawl, accessibility and liveability, social balance and other urban issues.

Europe and the Member States need to consider fully the potential of cities and towns and develop stronger, and perhaps more formal cooperation. The Covenant of the Mayors initiative is a step in the right direction. To be fully effective, cities need to combine their efforts with those of other cities and with national and European initiatives.

Box 2.19 UK adaptation of 19th century houses to climate change

Problem
Projections for climate change in the United Kingdom indicate that peak summer temperatures could be up to 7 °C warmer than today by the latter decades of this century, but buildings in the United Kingdom have evolved to provide thermal comfort in a temperate northern Europe climate. The following modelled case explores possibilities to adapt 19th century family houses, typical of many towns and cities in the United Kingdom.

Situation
The house as built has four bedrooms and is semi-detached. It has a brick and render façade and a slate roof. The building is poorly insulated, with solid wall construction and single glazed windows. Ventilation is provided by opening windows.

During the summer indoors, the 'warm' and 'hot' discomfort temperatures are increasingly exceeded for house interiors over the next decades, putting the rooms into a heat stress zone (Figure 2.24). The failure of the building to regulate indoor temperatures is a consequence of a number of factors but particularly the lack of shading from the sun and poor control of ventilation.

Solution
The adapted house has solar shading: external blinds or shutters capable of screening out 95 % of sunlight during the day, and ventilation: a secure means of ventilation, capable of providing ventilation rates similar to those provided by opening the windows. Here, it is assumed the ventilation system is mechanical, but it could potentially use natural ventilation. The ventilation system provides maximum ventilation whenever indoor temperatures are above 24 °C, and above the outside temperature.

Results
The adaptation measures considerably reduce the proportion of hours in which the discomfort temperatures are exceeded (Figure 2.24). However, the 1 % overheating limit is exceeded from the 2020s onwards in the bedroom (3 % exceedance) and from the 2050s onwards in the living room. The adaptation measures have a limited effect on reducing peak temperatures.

Figure 2.24 Discomfort temperature as built (left) and as adapted (right)

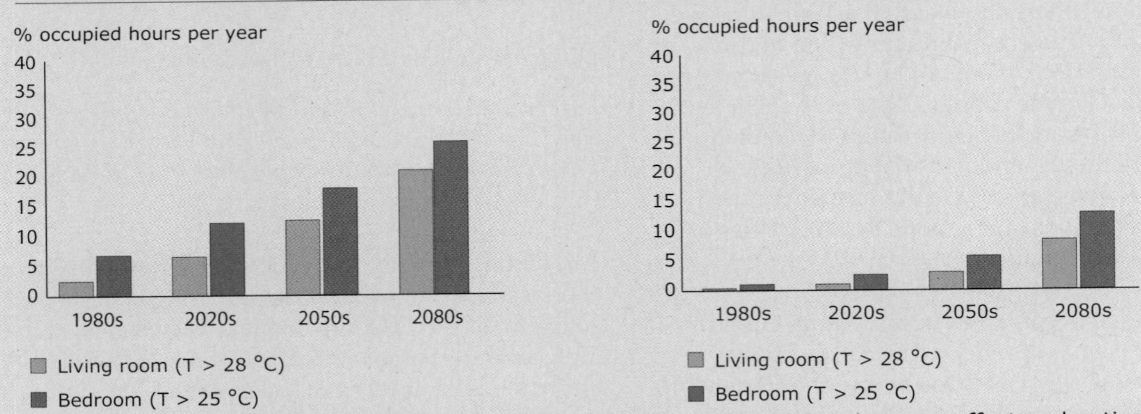

Since the adaptation measures are only applied during the summer they have no affect on heating energy. In summer, additional energy is required to power the fans for the ventilation system, but the predicted energy consumption of the fans is relatively small. The house's greenhouse gas emission is expected to decrease.

The house could alternatively be kept cool using air conditioning, but this increases the greenhouse gas emissions more than the reduction in emissions resulting from lower winter heating offsets.

Source: Hacker, J. N.; Belcher, S. E. and Connell, R. K., 2005; more information: www.ukcip.org.uk.

Box 2.20 The Netherlands — learning from bad and good practice

Bad practice — new settlements 1970–1990

Various new urban settlements in the Netherlands between 1970–1990 implemented ambitious sustainable building practices (DUBO), but they all neglected efforts to prevent the effects of flooding, drought, increased energy consumption for cooling, etc.

Some of these suburban areas have already been demolished and rebuilt due to social problems. In most cases the original ambitions of providing quality of life were never met due to a lack of money as well as lack of interest from developers and local public government for future end user needs. They wanted a quick return from investment and had no time for realizing sustainable solutions.

Lessons learnt
Integration of know how and knowledge in the decision-making process would be more effectively enabled by the active participation of stakeholders (learning by doing) instead of more instruments or new regulations.

Good practice — building with Water Haarlemmermeer

Problem
In the Haarlemmermeer polder area (Schiphol area), there is concurrently an increasing demand for housing and for water peak and seasonal storage in the same space.

Solution
The municipalities of the area developed an efficient and effective plan for multifunctional climate proof urban development, based on the storage of water. Public agencies at all levels, together with private agencies (building, developing and corporate housing) joined forces with various knowledge institutes throughout the Netherlands.

Results
The plan to designate a large part of the Haarlemmermeer polder (Schiphol area) for experimenting with adaptation principles achieved a broad acceptance. The approach also contributes to solving problems at regional and European scale by accumulated knowledge on adaptation towards flooding and droughts risks.

More information: www.bouwenmetwater.nl.

The European Commission should provide assistance to Member States in the development of national adaptation strategies with a local dimension, and a local role should be defined for the different actors at the local level, with special focus on small and medium-sized initiatives. The Commission can support cities and regions by providing tools, promoting information exchange and cooperative approaches and by helping them produce and implement action plans on climate change. These plans should include actions on both mitigation and adaptation, based on robust and clear objectives, and a defined time plan. One method of support is to provide better access to information on climate change mitigation and adaptation actions and good practice at city level. Information on adaptation actions could be provided by the proposed European Climate Change Impacts and Adaptation Clearinghouse.

Developing and distributing tools

Many tools can help in combating climate change and its effects. Some key elements include:

- Awareness-raising campaigns at the EU and local level to address citizens' attitudes towards both the mitigation and adaptation to climate change. These campaigns must provide very clear messages to citizens on their contributions to the reduction of emissions and their impacts, but also provide a clear message about the consequences if attitudes and behaviour remain unchanged. Indeed, this is common in many countries, such as the United Kingdom, where national agencies provide information and guidance on ways to reduce the carbon footprint, whilst local authorities work with partner agencies to provide information and services. An increased understanding of the context and science behind the climate change debate and the probable impacts of inaction could prove invaluable in gaining support for acceptance of potentially controversial measures (see also example in Box 2.16). Awareness raising and increase of knowledge is equally important among decision-makers at the city level if they are to get broad support for climate-friendly developments.

- Compulsory sustainability impact assessments at the EU level to help provide answers to questions concerning the future model of the sustainable urban environment. Given the need to reformulate urban planning, these assessments should be made with the climate change projections for 2050 in mind.

- Structural Funds checked against their impact on climate change. This is essential, and can be used to support cities' adaptation to climate change. Structural Funds will include specially designated adaptation funds at the local level, to help ensure that, for example, both new and existing buildings can be made climate neutral and climate proof.

- Regular and systematic involvement of cities and regions in discussions on how to integrate climate change adaptation measures into all policies. There is a need for more involvement of stakeholders as mentioned in the Commission Green Paper on adaptation to climate change (EC, 2007f).

- Improved knowledge base, in particular on the local and regional level information across Europe. Integrated climate change research, including research on micro-climates and urban heat island effects can also reduce the uncertainty with respect to adverse climate change impacts. Research and action should also focus more on the social and cultural dimension of the impacts of climate change.

- Exchange of good practice information, and in particular that concerning climate change adaptation, as well as discussion of the means to implement good practice (Box 2.20). The mainstreaming of such key exchanges through city networks and programmes like Interreg, as well as the proposed European Climate Change Impacts and Adaptation Clearinghouse can provide a robust platform for future work.

- Making transport policy a priority. There must be a shift towards more sustainable modes of transport, and to reducing transport needs by appropriate city planning and design.

2.6 Cohesion policy

In 2004, when ten new Member States joined the European Union, it was anticipated that the benefits of stability, higher economic growth and improved quality of life would be achieved by these countries, in the same way that recently Ireland, Portugal, Spain and Greece had benefited. These shining success stories of European integration over the last years transformed some of the continent's poorest members to wealthy economies. However, economic success has come at a price. The highest increases of greenhouse gases, doubling in the case of Spain between 1990 and 2006, accompanied this economic transformation, and so undermined the quality of life. Europe policies in the pursuit of the necessary economic growth and regional convergence must be based on a balanced approach, integrating social and environmental considerations and recognising that, in Europe at least, quality of life is mostly but not wholly dependent on higher incomes (Eurofound, 2007).

The preceding sections advocated the need for a strong integration of policy levels and actors. This section focuses on European cohesion policy as one illustration of the interdependencies between European and local policy and its likely impacts on quality of life. This chapter explores options to avoid the negative side effects of sustainable development and maximise the positive effects in order to improve quality of life in cities and regions across Europe in economic, social, cultural and environmental terms.

Overcoming disparities

Due to its historic development, Europe demonstrates economic, social and territorial disparities between its regions apart from natural disparities such as geographic or climate specifics. Today, development is also driven by global forces and enhanced competition between regions on a global scale, major demographic trends, and climate change. The Treaty of Rome (1957) recognised these differences and set out the vision that 'the Community shall aim at reducing the disparities between the levels of development of various regions'. Determined action backed by real resources only began with the creation of the European Regional Development Fund (ERDF) in 1975. Today the Fund operates in liaison with the European Social Fund (ESF); the two funds, together known as the EU Structural Funds, represent the major instruments of EU cohesion policy.

Over the years, many disparities between the regions have been eroded; although, some may have persisted, hidden within statistical averaging. However, the accession of the ten Member States in 2004 doubled the development gap between Europe's regions, and greatly increased the population subject to social exclusion. Most beneficiaries of cohesion policy are now located in central and eastern Europe, and these changes required a major overhaul of cohesion policy resulting in new guidelines for the period 2007–2013 (EC, 2006b) (Box 2.21). Cohesion policy also supports other sectoral EU policies to meet the aims of the Lisbon Strategy for jobs and economic growth while respecting the needs of the Gothenburg and the renewed Sustainable Development Strategy, which target the material basis of quality of life.

Urban dimensions

In 2006, Danuta Hübner, Commissioner for Regional Policy, said: 'Europe's towns and cities have a vital role to play.. They are the motors of growth and jobs and centres of innovation and the knowledge economy. At the same time, urban areas are the frontline in the battle for social cohesion and environmental sustainability.' (Hübner, 2006)

The Communication on cohesion policy and cities (EC, 2006c) stresses the importance of cities and towns. The current regulations applicable to the Structural Funds explicitly include the urban dimension and territorial cooperation. The aim is also to strengthen polycentric development in Europe and cross-border cooperation by promoting joint initiatives at the local and regional level, thus providing cities huge opportunities to develop sustainably. Although there are no programmes directly targeted at cities in the Structural Funds, the Operational Programmes can include such projects. Hence, for the period 2007–2013 the Operational Programmes funded by the ERDF have allocated the EUR 10 billion to Specific Priority Axis on urban development and many other projects indirectly related to urban areas (EC, 2008b).

Broader policy frameworks

EU cohesion policy should not stand in isolation but must form a major part of a complex combination of policies at all administrative levels. The Structural Funds provide financial support and leverage effects for regional economies. The main interventions affecting territorial development, in particular the urban environment, are shown in Table 1.2 and EC, 2007h. Successful implementation of cohesion policy depends on macroeconomic stability and structural reforms at national level together with a range of other conditions favouring investment (EC, 2006b). Cohesion policy can, therefore, only be effective with the full support of the legal systems affecting land-use and land planning, such as taxation — particularly taxation related to property — and zoning or land ownership registers at the local level. However, decisions on these policy areas remain the preserve of national, regional, or local authorities.

The implementation of cohesion policy is therefore the responsibility of all partners, in particular the managing authorities at national or regional levels. The EU acts in accordance with the Treaties, but the principle of subsidiarity guides community interventions, which requires that matters are

Box 2.21 Three pillars of EU Cohesion policy 2007–2013

Convergence
Supports the least developed member states and regions with more than 80 % of total expenditure, funding amongst others, projects in environment, risk prevention, energy, and transport.

Competitiveness and employment
Supports the more developed regions, funding projects including protection of the environment and risk prevention, for example the cleaning up of polluted areas, supporting energy efficiency, and clean public transport.

Territorial cooperation
Aims at cross border activities, transnational and inter-regional cooperation. Programmes funded via INTERREG and URBAN II support, for example, exchanges between cities on sustainable urban development (examples Boxes 2.24 and 3.3).

handled at the lowest competent level. Successful cohesion policy requires broad participation of stakeholders at all levels.

Cohesion policy and quality of life

Cohesion policy is one of the most powerful EU policies, deploying 35.7 % of the total EU budget for the period 2007–2013. The wide range of supported activities drives positive change and aim to enhance quality of life and the environment of cities and towns in Europe. Cohesion policy is at the very core of issues concerning sustainable development, as it aims to support economic and social development whilst safeguarding the environment by, for example, projects concerning urban transport and the revitalisation of city centres (Box 2.22). Approximately 30 % of the Structural Funds is allocated for environmental programmes. The decoupling of economic development and environment degradation is a fundamental challenge that must be met.

Notwithstanding the clear and positive objectives of cohesion policy, in some cases projects supported by EU Structural Funds may cause unintended side-effects. For example, measures to increase accessibility, which cities and towns can benefit from, also lead to increases in transport demand and exacerbate problems of noise, air pollution,

Box 2.22 Wrocław (Poland) — Structural Funds supporting public transport projects

Situation

Public transport still faces major problems in Europe. The transport policy of the new EU member states, in particular, focuses strongly on road infrastructure, and in many of these countries the national government has transferred responsibility for public transport to municipalities. National or regional funds in public transport, such as those in Germany, are usually not available. As a consequence, important imbalances remain among cities and regions across Europe in terms of the financial resources to provide high quality public transport and its availability.

Photo: © Krystyna Haladyn

Another problem concerns the fact that effective public transport concepts often require much greater coordination than road projects. As a result, the implementation of such concepts is often delayed or frozen. In both contexts, support from EU funds is crucial in giving the necessary impulse, in particular in cities and regions that are lagging behind in both old and new Member states.

Solution

The city of Wrocław, Poland, faced such problems when it, together with environmental groups, started to develop a concept to improve and increase the attractiveness of its public transport system. This concept included a tramline expansion and the purchase of new vehicles as a pilot project, but implementation failed because of a lack of resources. However, Wrocław and its partners further developed the concept and implemented many smaller, low-cost measures.

Results

The low-cost measures have already led to an improvement of the public transport system of the city and its neighbouring municipalities. Moreover, the integrated concept finally convinced European institutions to dedicate EU funding to Wrocław in order to realize this and a further tram line. This EU support will help to increase further the attractiveness of the city's public transport system and will offer a real alternative to car travel that can contribute to limiting and reducing negative environmental effects from transport.

More information: www.wroclaw.pl.

additional land take and fragmentation, and climate change (Box 2.23). Possible side effects must be considered in evaluations of the overall effectiveness of cohesion policy.

Avoiding negative impacts

Operational Programmes (10) form the basis for requesting Structural Fund support. Through these

Box 2.23 European cohesion and transport policy — improving regional competitiveness with unintended side effects

Improving the accessibility of regions and cities through the European Cohesion and transport policy, as well as through national and local transport policies is seen as a key factor underlying regional competitiveness and growth. Structural and Cohesion Funds have been used intensively to improve accessibility and support the development of transport infrastructures.

For the period 2007–2013, more than EUR 80 billion are allocated for transport in the Structural Funds. 51 % of this budget is foreseen for road and air transport projects, 47 % for sustainable modes — rail, shipping, cycling, multimodal and intelligent transport systems and clean urban transport, 2 % for urban transport. Two thirds of the EU-27 budget is allocated for the 12 new Member States where transport is given a high priority amounting to 20–38 % of the Structural Funds. In most new Member States road projects are prioritized highest; meanwhile in many older Member States, sustainable transport projects have a higher priority, as many road projects were completed in the last period.

Greater accessibility typically increases movement between cities and enables wider commuting areas, thus increasing demand for transport. Depending on the mode of transport, it can involve more traffic congestion, air pollution, noise, impacts on human health and safety, and support urban sprawl, which further increases the demand for road transport in particular (see case study in Box 2.9). It will also contribute further to the emission of greenhouse gases and climate change effects, and reduce areas for biodiversity and ecosystem services upon which our quality of life and future development depend.

Furthermore, unbalanced project implementation, for example by prioritizing road extension over train infrastructure, can lead to an overall unbalanced development increasing the negative effects of transport. The economy, territorial development and built environment, especially of the new Member States, will orient their logistics and infrastructure towards the road infrastructure that is already available. A later shift back to rail will be costly and complicated, especially if it is to be of good quality. The increased share of road transport will therefore contribute even more to noise and air quality problems in cities and their environs and can counteract local activities to promote public and non-motorized transport.

Figure 2.25 Distribution of the Structural Funds 2007–2013 allocated to transport

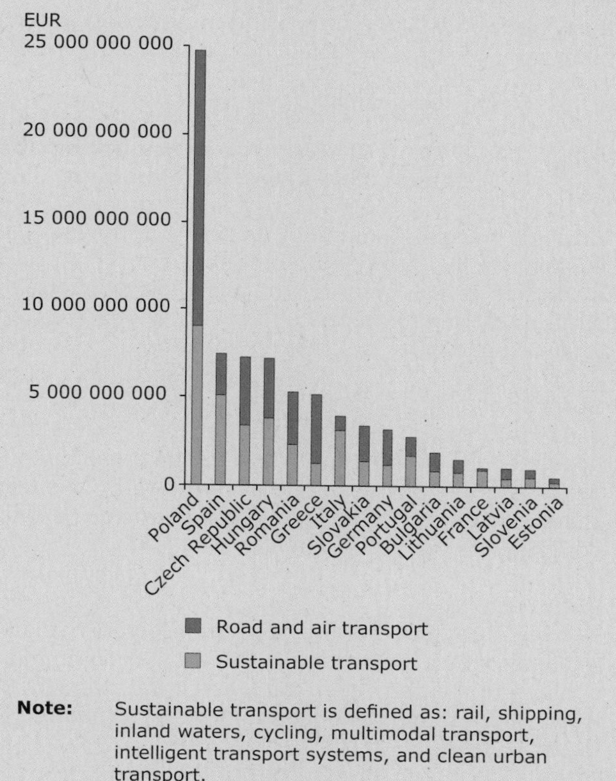

Note: Sustainable transport is defined as: rail, shipping, inland waters, cycling, multimodal transport, intelligent transport systems, and clean urban transport.
Austria, Belgium, Cyprus, Denmark, Finland, Ireland, Malta, the Netherlands, Sweden and United Kingdom are not shown due to only marginal fund allocations.

Source: DG Regio, 2009.

(10) For a state of play of the submission and approval of National Strategy Frameworks and programmes, see http://ec.europa.eu/regional_policy/index_en.htm.

funds, regions are able to support projects in cities and towns. The Community Strategic Guidelines 2007–2013 (EC, 2006b) request an integrated approach and imply compliance with the precautionary principle, efficient use of natural resources and the minimisation of waste and pollution, thus including quality of life in all its dimensions. The Commission guidelines on cohesion and cities (EC, 2006c; EC, 2007h) provide further urban-specific guidance on these issues. In addition, the EU Directives on Environmental Impact Assessment and Strategic Impact Assessment require the assessment and minimisation of any potential negative environmental impacts of the projects, plans and programmes.

As environmental projects are included under different expenditure chapters of cohesion policy, a number of studies have explored the effectiveness of environmental interventions financed by the Structural Funds (EEA, 2009b) and have identified key issues, including the importance of the horizontal perspective of environment policies and their integration in sectoral policy interventions such as the Operational Programmes and investment projects financed by Structural Funds. This is particularly relevant for urban development, as the effective combination of the different dimensions of sustainable development is a major challenge.

Past Community initiatives have also supported the promotion of sustainable urban development, including the URBAN initiative, URBACT, INTERREG, ESPON, Leader+ and Equal. The URBAN initiative is particularly noteworthy, as it aimed specifically to tackle urban areas in crisis and promoted integrated and partnership approaches not only increasing and distributing knowledge but also cooperation among different stakeholders instead of competition (Box 2.24 and Box 3.3).

Barriers to policy implementation

The strong engagement of cohesion policy with other policies and actions at all administrative levels obviously requires an integrated approach. This should apply to the development of the Operational Programmes and to the individual projects although often this does not occur in practice. For example,

Box 2.24 URBAN II Initiative in Mannheim and Ludwigshafen (Germany) — cooperative urban development

Urban II was the Community Initiative of the European Regional Development Fund (ERDF) for sustainable development in urban districts that promoted the design and implementation of innovative models of development for the economic and social regeneration of urban areas. It played a role in strengthening information and experience-sharing on sustainable urban development in the European Union, in particular by creating the URBACT network.

Situation
Ludwigshafen and Mannheim are at the centre of the Rhine-Neckar Triangle, Germany — a densely populated urban region with 2.4 million inhabitants. The Rhine River separates the cities and also forms the border between two federal states, yet Mannheim and Ludwigshafen are interlinked due to their spatial proximity. During the 1990s both cities experienced negative economic development and declining population.

Solution
To overcome the declining trend Ludwigshafen and Mannheim pooled their efforts and started a cross regional project under the EU URBAN II programme, supported financially by Structural Funds. In addition to the two cities, the two federal states of Baden-Wuerttemberg and Rheinland-Pfalz and other local, regional and national authorities coorperated in the broad participatory approach of the project (2000–2006).

The project focused on a comprehensive approach across all policy areas — economic, social, cultural and environmental — in order to support integrated sustainable urban rehabilitation. Good practice, developed in several pilot projects, should spread over the entire region, become the normal way of conducting business and trigger an ongoing revitalization process.

Box 2.24 URBAN II Initiative in Mannheim and Ludwigshafen (Germany) — cooperative urban development (cont.)

Photo: © Stadt Mannheim

Examples of such projects include:

- targeted support to small and medium enterprises thus creating new jobs;
- clean-up of inner-city brownfield areas as sites for business and other uses;
- rehabilitation of pedestrian zones and creation of new high quality public spaces along the rivers increasing their attractiveness for people and thus for business activities;
- many small social projects supporting the integration of disadvantaged social groups

Results

Despite the fact that it is too early for the project to show any reversal of the overall trends, the actions that have been initiated have shown early effects, in stopping the downward trend and increasing the attractiveness of the cities, and have even led to a slight increase in population. The public information activities also supported people's awareness and identification with their city, and increased their voluntary engagement. This has already enabled some of the small projects, like an internet café for pensioners in Ludwigshafen or two cultural events in Mannheim, to operate further with their own resources.

Ludwigshafen and Mannheim have cooperated very closely in the economic area, against the trend of competing regions and cities. The joint approach avoided wasting financial resources in competition and led to benefits for both cities. The URBAN project even stimulated further cooperation in other areas like construction and infrastructure.

The URBAN project enabled the testing of innovative approaches. For instance, it increased awareness of the important effects that can be achieved by small social projects. The cities had developed these projects consciously to enable citizens' active participation and to support larger projects with higher investments. The will to cooperate, openness and a comprehensive participatory approach were the important factors in the success of this project.

More information: http://www.ludwigshafen.de/standort/3/urban_ii/, www.mannheim.de.

particularly in the new Member States, Operational Programmes include sectoral urban projects rather than integrated 'URBAN II like' projects (EC, 2008b). The example of wastewater treatment (Box 2.25) demonstrates the potential benefits in the form of effective as well as efficient solutions that can be delivered through more coordinated actions. One problem to date is that no common or precise definition, guidance or standards for the integrated approach exists, and it is up to individual countries and regions to define their own approach including the level of integration and the participation of relevant stakeholders. Poor planning and implementation of programmes and projects at local and other levels can result in ineffectiveness and low uptake of public funds.

In spite of their important role, cities and other local stakeholders are not necessarily directly involved in the development of the Operational Programmes, though many projects concern urban issues. The level of stakeholder involvement depends on administrative practices and processes in place, and varies between Member States.

Some projects require the involvement of only a few cities and some the involvement of many. Coordinating those involved can be complicated and can delay projects, leading to imbalance in the implementation of Operational Programmes (Box 2.23).

In assessing the environmental impact of projects, instruments such as Environmental Impact Assessment and Strategic Impact Assessment are available, but again, there are insufficient comparable standards at the community level to ensure high environmental standards following project implementation in all Member States.

Plans, programmes and projects eligible for cohesion funding must comply in principle with EU environmental legislation (cross-compliance principle). Apparent contradictions are not the result of cohesion policy, rather the causes lie primarily in EU environmental legislation itself, and the way it is implemented and enforced by the Member States and supervised by the Commission. Another cause lies in the absence of clear environmental conditions in the cohesion framework (Box 2.26).

The actual impacts of cohesion policy and projects on other than the target areas; for example, by contributing to urban sprawl or urban and regional transport growth, are only partially understood, which hinders the adaptation of policies to minimise these adverse impacts.

Overcoming barriers to action

When national progress is evaluated, social and environmental indicators should be considered alongside economic indicators. A recent survey showed that 67 % of people in Europe think that progress should be measured in terms of these objective elements of quality of life (Eurobarometer, 2008a).

EU cohesion policy takes into account the fact that in the long term high income and economic growth alone do not ensure a socially balanced quality of life, for which a healthy environment and enhanced cultural and societal values are equally important. This highly challenging balancing of priorities requires further discussion and innovative responses, along with further insight into the issues associated with territorial cohesion — a debate rekindled by the Commission's Green Paper on territorial cohesion (EC, 2008c) and the Beyond GDP conference and initiative. For example, is it appropriate to focus on growth, in particular if measured in GDP terms, in every region given the legacy of historically diverse development in Europe and the reality of an ageing and declining population? How can we turn the diversity of Europe's regions into an asset and how can we ensure that cities and regions in Europe collaborate to tackle current and future challenges, instead of competing with each other?

Bearing in mind the many interlinkages and the fact that problems like climate change or urban sprawl cannot be solved at one administrative level alone, the stakeholders of cohesion policy implementation need to develop further and apply effectively an integrated approach.

EU and integrated approaches

The European Commission needs to integrate its policy areas and support administrations in the Member States so that they can fulfil their responsibilities for integrated action. With respect to cohesion policy, this requires the development of better guidance on how to formulate an integrated approach that takes into account the EU Territorial Agenda and the *Leipzig Charter on sustainable European cities*. The Commission should further promote the application of such guidelines and, as far as possible, strengthen the development of an integrated approach in regions and cities when granting Structural Funds, as requested recently by the European Parliament in its report on the Territorial Agenda (European Parliament, 2008) (see also Chapter 3). Creating awareness amongst regional and local stakeholders based on the aims of the

Box 2.25 Integrated action to improve the efficiency of Structural Funds

The case of urban wastewater treatment

The urban wastewater sector had the largest share of the allocation to the environment through the Structural and Cohesion Funds (ENEA 2006). The discharge of urban wastewater into rivers, lakes and the sea is a matter of great concern in most countries. The EU Urban Waste Water Treatment (UWWT) Directive (Council Directive 91/271/EEC) requires that all communities above a certain size install adequate collection, treatment and sludge management systems to dispose safely of the urban wastewater they generate.

However, despite three decades of effort only about 54 % of EU-15 cities complied to the wastewater treatment levels required by the Directive. Cohesion Funds can help to close this gap but need to be applied efficiently.

Land use and wastewater treatment

Urban population and urban sprawl are important factors when planning the development and positioning of urban water treatment plants. For example, in Spain about 55 % of the population is connected to urban wastewater treatment plants, with the lowest connection rates in coastal areas (EEA, 2005). Meanwhile the map of urban sprawl shows that urbanisation had primarily taken place in these coastal areas. This raises the question as to whether fewer people will be connected to the UWTP at the end of 2006 than in 2000.

Map 2.17 Distribution of Cohesion Funds spent in sewerage and purification compared to urban sprawl in Spain

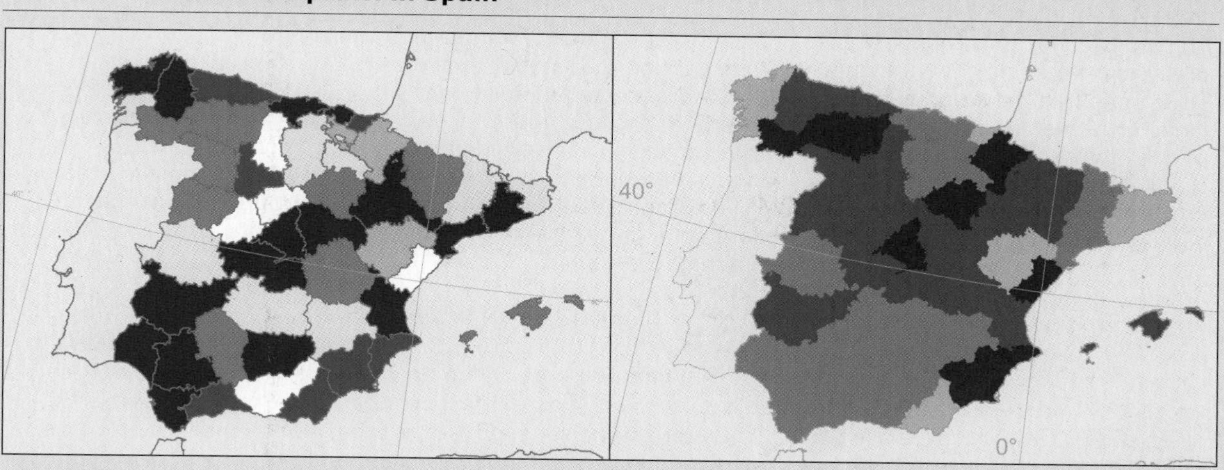

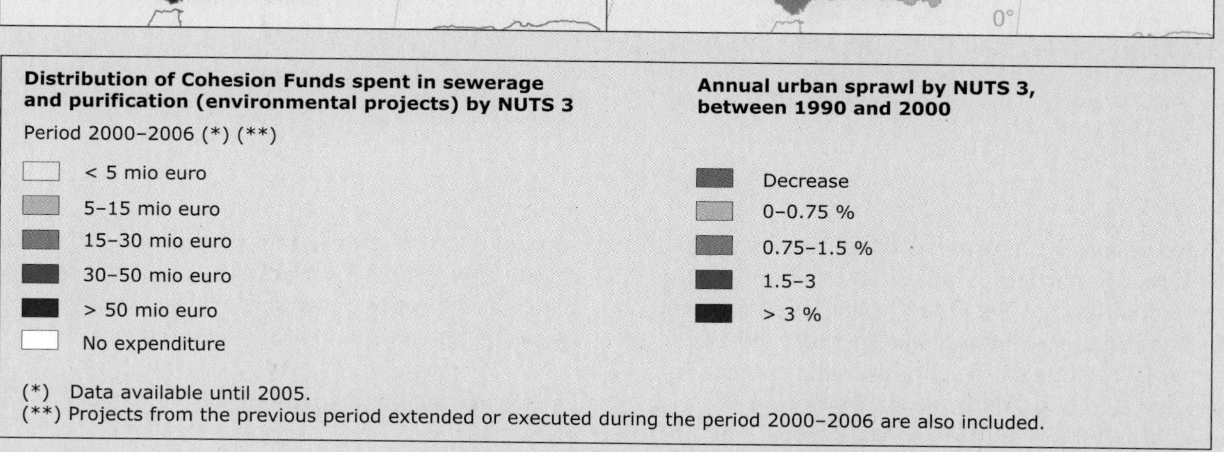

Source: EEA, 2005.

Box 2.25 Integrated action to improve the efficiency of Structural Funds (cont.)

Complementary economic instruments and incentives
The results of an EEA study (EEA, 2005) for some European countries indicate that there is a risk of excessive investment in sewage-treatment capacity in the absence of complementary economic instruments to provide industries with an incentive to promote eco-efficiency and to reduce pollution at source. For example, the share of population served by public treatment in Spain was relatively low compared to the total expenditure. In contrast, the Netherlands showed that the use of economic instruments as an incentive to industry to reduce discharges at source has reduced the need for public sewage-treatment plant capacity, and public investment, to a level well below that in other countries.

Conclusion
All these results contain interesting signals as to how to make Cohesion Funds in wastewater treatment services even more effective. The data and results in both cases should be further analysed. A broad integrated approach is absolutely necessary.

Source: EEA, 2005.

Box 2.26 Transport and cross-compliance principles

The cross-compliance principle means that the Operational Programmes concerning transport should undergo a strategic environmental assessment and that most transport projects should undergo an environmental impact assessment. However, these regulations do not set any measurable environmental limits or targets and mostly focus on procedural aspects, which leave a large margin for manoeuver by the Member States and their authorities regarding the selection of the mitigation measures. Unfortunately, these laws are often seen as a bureaucratic exercise rather than a tool to deliver a better environmental outcome. Furthermore, EU environmental legislation does not set targets and limits directly applicable to the noise and air emissions brought about by transport plans and projects. The setting of such standards is at the discretion of the Member States.

An unsustainable transport plan or project may therefore be given consent without breaching EU law in a heavily polluted area. For instance, if the emissions arising from the plan or project are anticipated to cause exceedance of EU air quality limits or to increase pollution in areas where the limits are already exceeded. In the absence of EU standards on ambient noise, transport plans and projects may also be given consent without breaching EU law on ambient noise, even if they bring about noise levels above WHO guidelines. Implementation of EU environmental laws thus raises a number of concerns which hamper the efficient control of environmental outcomes by the Commission.

guidelines for Structural Funds (Hübner, 2008a) and encouraging participation in the development and implementation of the Operational Programmes can help maximise overall benefits and minimise negative side effects. The integrated approaches developed in URBAN projects have proved successful but need further promotion if they are to become mainstream.

The Commission will also need to analyse the effectiveness of current instruments, such as Environmental Impact Assessment and Strategic Impact Assessment in order to avoid negative effects on the environment and must support the Member States in their move towards a more homogeneous implementation.

Cities and participation

To benefit from cohesion policy and Structural Funds cities and towns must participate in the process of elaborating the Operational Programmes at the regional level. They must ensure that their projects are integrated into a carefully planned sustainable development concept and future vision for the city or region. Their participation should comprise not only measures that address the city or town directly, but also proposed measures that might have indirect impact. For example, similar projects in nearby cities that could increase competition and hence will lower the success rate (see Box 1.9, bad practice and Boxes 1.10 and 2.27, good practice) or European and cross-regional transport projects, which can substantially change the local situation. Box 3.5 on the *Magistrale für Europa* initiative, which can be found later in this report, provides another good example of cities' participation in European Transport policy and its implementation. This is also a good example of the success of cohesion policy. Active participation is rewarded by more sustainable and balanced development to the benefit of the majority. The newly created instrument of European Grouping for Territorial Cooperation might provide an appropriate tool for better participation.

Enabling participation

Integration and a participatory approach based on a sustainable development strategy at the regional level is a key factor of long-term success. The regions have considerable responsibility for enabling their cities and towns to improve their situation and minimise disparities. Regions should also make sure that the implementation of cohesion policy meets all the requirements of sustainable development, that assessment tools are applied appropriately, and should enable broad participation of all relevant stakeholders.

Regulations alone, at whatever level, are not guarantees of success. Good work by the managing and certifying authorities is crucial to ensure correct policy implementation and use of taxpayers' money. If not, subsequent audits by the authorities and the Commission will not indicate satisfactory outcomes (Hübner, 2008b).

Improving knowledge

Assessing the effects of cohesion policy in urban areas, both positive and negative ones, is a complex task, and it is not always possible to identify cause–effect relationships. Because of this, research programmes and knowledge exchange should be targeted towards closing information gaps, in particular those relating to unintended impacts. The Commission needs to find ways anticipating what these effects may be. This will need to be done in particular at high level as it would appear that most questions cannot be answered at regional level alone.

An improved knowledge base across Europe would also form the basis for the spatial approach required if territorial cohesion is to respond more effectively to the specific territorial needs and characteristics, geographical challenges and opportunities of regions and cities. A more robust 'urban approach' would enable EU cohesion policy, as well as EU policy in general, not only to support cities via urban projects but also help to assess the likely impact of other projects.

Box 2.27 Czech Republic — improved integration of urban issues in the Operational Programmes

Initial situation
Cities need to participate in the Regional Operational Programme for the new period 2007–2013 in order to increase the take up of Structural Funds in these regions.

Solution
The Healthy Cities network of the Czech Republic, together with the Jihomoravský and Vysočina regions organized a series of local forums in the respective cities within the framework of the Regional Operational Programme. During these forums, local partners, including more than 1 200 people in 18 cities and their administrative territories — politicians and decision-makers, citizens, businesses, NGOs etc. — proposed those sustainable development projects that they wish to implement, discussed how these fit into local community strategy, identified and agreed on the priority projects and adapted local community strategies accordingly. Several local partnerships were established in the course of the local forums that will serve as platforms for further discussion in the future.

Photo: © HCCZ

Results
The priority projects selected were used to influence the future Regional Operational Programme so that its priorities best reflect local needs. According to the consultations, the participants would like to invest in new infrastructure, mainly roads, water treatment plants or in the area of tourism. More than 2000 project proposals were collected from all partners, both public and private, in the Vysočina region and the regional office is still working with these proposals. To accommodate and to analyze such a huge number of proposals, the Vysočina region used an Internet information tool DataPlan provided by the Healthy Cities association. All information is therefore accessible to the public and can be used for further procedures according to actual needs — connecting with budgets, creating more complex projects of regional importance etc.

Despite the many hard outputs, soft results were also achieved. Vysočina succeeded to create communication channels between various partners, local governments and the regional government. The communication channels are still in use, for instance for the implementation of successful projects. Mrs. Marie Cerná, former deputy chief executive officer and HCCZ vice chairman, concluded that long-term dialogue between those who have concrete ideas about the development of their living space and those who are able to fulfil these ideas was also established, owing to the fact that the key principle of partnership was filled with real content.

More information: www.nszm.cz.

3 Towards integrated urban management

'Partnership between the local, regional, national and European levels of government will ensure we can cope better with common global challenges ahead,' said Luc Van den Brande, President of the Committee of the Regions at the Brussels Open Days of the European Regions and Municipalities 2008.

Most of Europe's population living in major metropolitan regions is at risk. City dwellers have high expectations, and most expect their lives to be more pleasant over the next five years (Eurobarometer, 2005). But the current patterns of urbanisation and forms of most new urban development are unsustainable and becoming increasingly so, putting at risk the quality of life of inhabitants. These conflicts and tensions in urban development can become a source of increasing pressures on city governments to deliver a better way forward. The integrated and collective response to urban governance provides the potential to reverse these trends. If the EU is to tackle these issues, and in particular the over-riding challenge of climate change, then it must increasingly be an active partner in the governance of Europe's towns and cities.

Building on the outcomes of the previous chapters, this chapter summarises the needs for an integrated urban approach and provides ideas about how to develop and implement it collaboratively across all administrative levels.

3.1 EU and cities partnership

The preceding chapters highlighted some of the important challenges facing Europe's cities and towns in securing a long-term and socially balanced quality of life. Local city-based programmes, policies and projects remain key to delivering the required action, and numerous local initiatives demonstrate that European urban areas are already strongly committed to the need to improve the quality of life of Europe's towns and cities.

Urban areas have the responsibility to regulate and manage urban policy and effective planning strategies in the interests of the local population; however, no city is self-contained. Urban Europe is a mosaic of overlapping and complex polycentric metropolitan regions in which context urban development is driven and guided at all government levels as described in the chapters before.

The European ideal is based upon the central concept of a common future. The Lisbon Treaty builds on this concept and has reinforced a culture of cooperation and integration between governments and their communities. This vision of cohesion and cooperation has never been more essential or urgent than now in meeting the many challenges facing cities, including globalisation, the need to secure sustainable energy sources, the impacts of demographic shifts, as well as the growing threats of climate change and to national security. Urban areas are also central to EU economic and social policies and programmes as key drivers of economic growth (EC, 2006c).

Table 1.2 in Chapter 1 of this report identifies a selection of the large number of relationships that exist between European and local policy in different policy areas and shows where financial resources and other incentives steer urban development. In particular the Structural Funds of EU cohesion policy have had and will continue to have, major direct and indirect impacts on urban development. Also, the implementation of the Trans-European Transport Networks (TEN-T) has been key in redefining the relationships between the cities of Europe, the patterns of movement, logistical systems and economic activity. European policy together with the policies of the member states and regions, provide the framework and general conditions for the realisation of quality of life in cities and towns. Cities and towns implement measures on the ground and create the conditions for quality of life and sustainable development.

In a globalising world, cities and towns in once peripheral regions are becoming increasingly accessible, and locational choices, including those for new urban investments, are generally more inter-changeable. As a consequence the scale and scope of action required is no longer the responsibility of any single sector or level of government. Furthermore, European integration has not simply shifted authority upwards to European institutions; rather authority has become increasingly dispersed through a variety of different levels, actors and agencies, creating a multilevel basis for governance (Rosamund, 2004). As a result cities seek to reinforce action at the local level by engagement in wider city regional networks and directly at the European level.

Nonetheless, the challenge remains to overcome isolated action at the local level and competition between cities and between regions by collaboration and integration from local to European level to the long-term benefit of all.

3.2 Integration gaps

Policy-making needs to reflect and respond to the many interconnections that lie in the fundamental drivers of urban development, yet the reality is that major gaps still need to be filled including:

- ***between sectoral policies***
  typically a plethora of plans and strategies exist for major urban areas, concerning transport, housing, environment, economic development etc. Policies within these different documents are often based on different assumptions and timescales, and with no regard for the unintended impacts on other policy fields;
- ***between plan-making and implementation:***
  the power to implement plans often lies with other agencies, and increasingly the private corporate sector. The challenge is to achieve the integration of plans and programmes and to engage all stakeholder interests, including small business and the community, without compromising effective implementation;
- ***between resources needed and available:***
  this particularly applies where major new infrastructure is required, for example transportation systems. The frequently high levels of new investment required for major projects often distort the political debate and create perceptions of unequal distribution of benefits between competing cities and regions;
- ***between administrations and functional urban regions:***
  few urban administrative areas relate effectively to travel to work, or labour market areas, or indeed natural regions. As a result urban and rural areas are frequently planned in isolation, and the associated competition between municipalities generates a resistance to collaborate on the development of the necessary common policy framework. This reluctance to collaborate is reinforced by the perception that economic growth merely diverts or displaces growth between urban areas. There are, however, excellent examples that demonstrate the benefits of joint working to achieve policy integration to learn from — see Stuttgart Box 3.1.

3.3 Barriers

Effective urban policy demands an integrated approach as endorsed in different documents ([11]). In some cases specific guidance is given, as with the EC Guide *Integrated environmental management* (EC, 2007g) that supports the implementation of the Thematic Strategy on the Urban Environment, but all too frequently such guidelines and in particular concrete criteria are not developed.

Local governments have developed integrated management approaches to improve consistency and coherence between policies also supported by a variety of EU-funded programmes ([12]). The many municipalities that signed the Aalborg Commitments ([13]) address all dimensions of sustainability using the framework of the Commitments and an integrated management for the implementation of local sustainability (see Box 3.2).

Despite the widely recognised need, general commitments and the availability of many tools and good practice examples, the reality is that integrated management across Europe is still a matter for a few pioneers. Isolated policy and individual interests still threaten sustainable development and longer term quality of life. Integrated management needs

([11]) Territorial Agenda, the Leipzig Charter on Sustainable European Cities, the Cohesion Policy guidelines 2007–2013, the Thematic Strategy on the Urban Environment and many others.
([12]) Including Managing Urban Europe-25, European ecoBudget, localsustainability.eu, Liveable Cities, and Dogme 2000 (localsustainability.eu; www.localmanagement21.eu; www.ecobudget.com).
([13]) See www.aalborgplus10.dk.

Box 3.1 Greater Stuttgart Region (Germany) — tackling integration gaps between city and region

Initial situation

The greater Stuttgart region has 2.7 million inhabitants and is the centre of industrial science and research organisations in Germany. To sustainably maintain its competitive status, it was necessary to adopt an integrated approach to the development of the 179 independent municipalities that made up the region.

Solution

The Verband Stuttgart was, therefore, founded in 1994 with 93 directly elected representatives in the Regional Assembly and an annual budget of EUR 260 million.

The range of joint responsibilities undertaken by the Verband Region Stuttgart included:

- a 10–15-year Regional Plan
- business promotion and tourism marketing
- transport planning and investment
- landscape and parks
- large infrastructure and investment (e.g. Paris-Munich high-speed train)
- waste disposal

Photo: © Stuttgart-Marketing GmbH

In addition the Verband can take on other tasks voluntarily, such as trade fairs and exhibitions. Some examples of joint actions include:

- Landscape planning and parks: the Verband created the 'Greater Stuttgart Landscape Park', showing where open areas are to be improved, redesigned, and linked together. The combined commitment of the Region, the municipalities, and all the various authorities is necessary to implement these plans.
- Traffic and transport planning: the traffic programme represents a blueprint for county and municipal planning and will ensure that the Verband is able to influence the investment programmes of the State of Baden-Württemberg and the German Federal Government. 85 % of its budget is devoted to local public transport. The region of the Verband is 'buying in' transport services from transport companies, such as suburban electric services from Deutsche Bahn AG (German Railways).
- Waste disposal: along with the rural districts and the City of Stuttgart, the region of the Verband is responsible for a segment of waste management (dump category II). In 1997, the Verband established standardized conditions across the region for waste disposal, leading to a considerable reduction in charges.

Results, lessons learnt and transfer potential

The lessons of this joint regional governance for the greater Stuttgart region have demonstrated the importance of providing a unified picture inwardly as well as to the outside world. There have been many direct outcomes. For example, for the first time, the region now has an integrated traffic and transport concept allowing buses to become part of an 'extension' of the suburban electric railway network and 24/7 timetabling. Similarly, whereas the region of the Verband was only able to plan the green infrastructure in the past, it now invests in specific projects together with local authority partners, providing a network of open spaces, ecologically valuable green areas and small parks combined with landscapes.

As a result of long-term cooperation and the joint and integrated approach, the implementation of measures became more effective and efficient e.g. the cooperation enabled larger and more complex infrastructure projects, avoiding competing measures which would have led to a waste of resources, and thereby increasing the attractiveness and quality of life for the whole region.

More information: http://www.region-stuttgart.org.

Box 3.2 Växjö (Sweden) — sustainable energy award, a successful integrated approach

Context of the municipality and initial situation

During the 1960's, two lakes of the Swedish City of Växjö became seriously polluted. A huge lake restoration project was launched in the 1970's and since then, many other measures have been taken in favour of the environment.

Photo: © Mats Samuelsen

The case: fossil fuel free Växjö through ecoBUDGET

In Växjö integrated and cyclic management is designed to achieve high environmental standards. In particular, 'Fossil Fuel Växjö' is an overall community programme that takes an integrated and cooperative approach to achieving its objectives. These include a wide range of activities aimed at generating more energy and heat from renewable energy sources and technology, improving energy efficiency in all areas, and achieving sustainable patterns of mobility. A major activity was replacing oil in the municipal district heating system with biomass — wood waste from the local forest industry. More than 90 % of the heat used in Växjö comes from district heating, and the network has been extended to outlying villages.

In 2001, Växjö began to introduce the integrated management system ecoBUDGET to ensure environmental improvement and efficient working arrangements. Through the system, the city can control the environmental resources in the municipality and monitor the implementation of goals in the Environmental Programme and the financial system. Within the ecoBUDGET framework, Växjö's political boards legitimize ambitious time-related targets supporting their objective to become fossil fuel free.

Results, lessons learnt and transfer potential

By 2006, CO_2 emissions per capita were already reduced by 30 % compared to 1993 — 3.5 tonnes of CO_2 annually. The biggest reduction in carbon dioxide emissions has been achieved by replacing oil in the municipal district heating system. The goal is to reduce fossil fuel carbon dioxide emissions per inhabitant, by at least 50 % by the year 2010 and by at least 70 % by the year 2025, compared to 1993. The city won the Sustainable Energy Europe Award in 2007 for its environmental efforts.

Växjö has become a role model, both nationally and internationally, for those seeking proof that sustainability pays. Collective environmental thinking over the last few decades has resulted in economic profits as well as cleaner air and water. According to Växjö officials, the municipality is well on its way to further achievements.

More information: www.vaxjo.se/english, http://ecobudget.com/.

to become mainstream across Europe. Transforming best practice in some municipalities to better practice everywhere requires keen examination of the barriers to wider exploitation of existing skills in, and experience of, implementation of integrated urban management. Examples of these barriers are given below.

- Even if overall sustainable development strategies based on an integrative concept are in place, sectoral and vested interests remain dominant where decision-making, administration and budgets are fragmented (lacking institutional integration) and decision-makers are not aware of the benefits of an integrated approach (see also Box 2.11). In general, governments are free to apply integrated management, and there are no penalties for failing to implement an integrated management system.

- Despite all the benefits that have arisen from EU policies and programmes, it must be recognised that these have been mostly sectoral in nature and project driven. More effective application requires more integrated — both horizontally and vertically — and comprehensive approaches to systematically address the common challenges; there is a need for 'urban-proofing' of policies and programmes.

- There is an over-riding governance deficit in the development of systematic approaches to EU policy to improve the management of towns and cities. The common challenges to the quality of life in towns and cities are increasingly beyond the control of local agents alone. Nonetheless, it is fully understood that the EU has no direct mandate for urban affairs, and its involvement in urban affairs must always be sensitive to the subsidiarity principle. However, the performance and development of cities and towns clearly has a European dimension, which must be addressed with supportive action.

- At the same time, it is also evident that cities and towns tend to resist greater engagement in local affairs from European and national levels. Subsidiarity requires that decision-making is undertaken at the lowest appropriate level. This risk of excessive parochialism needs to be recognised since there is not a single uniquely appropriate level for decision-making as most issues are linked via other levels and sectors. An emphasis on localism needs to recognise the risk of a 'democratic deficit' in society whereby those who are affected by decisions are excluded by administrative geography from those decisions.

- Existing spatial legislation can generally provide the basis for an integrated approach. However, the unsustainable development of the majority of urban areas demonstrates that planning legislation focusing on a traditional planning approach alone is often insufficient. The current legal planning system is mostly not suited to deal with the wide range of sustainability issues evident today. It cannot sufficiently take into account the rapidly changing environment and the need to adapt plans and the planning system so they are more comprehensive and spatially sensitive, embracing cyclical, integrated, inclusive and participatory approaches.

- Even where policy documents like the Strategic Guidelines for Structural Funds (EC, 2006b) request an integrated approach, it remains too often unclear what exactly is expected. No common standards exist, and at best only recommendations exist, to assist policy and decision-makers to define minimum criteria for an integrated approach.

- Socio-economic and geospatial data describing the existing state of urban areas are collected by municipalities and at higher administrative levels but this information remains sectorally specific. Sectorally specific formats differ, time series and spatial units inhibit effective application in the description and analysis of the urban system, how it is driven and the impacts of different trends and policies.

- Municipal networks including EUROCITIES, ICLEI, METREX, Energie-Cités, CEMR, and the Union of Baltic Cities are active with their member cities in the development of innovative approaches to the sustainable development of cities. Whether cities and municipal networks collaborate or compete depends amongst other things on both national and EU policy. When municipalities can apply for funds independently they tend to compete. On the other side, if national government or European Union funding permits collaborative action, they tend to cooperate and integrate (Kern & Bulkeley, 2009), as witnessed by the European Sustainable Cities and Towns Campaign, and projects in the context of the Interreg, URBACT, URBAN programmes. A good example is described in Box 3.3, another one earlier in Box 2.24.

3.4 Integrated urban management defined

The management of urban issues is complex and is influenced by a multitude of issues and stakeholder interests. The following description of the integrated approach to urban management provides general criteria applicable to all administrative levels, from local to European. Application in practice requires specific tailoring to individual and thematic circumstances.

Considerations for integrated urban management:

- **Urban** (towns, cities, conurbations, metropolitan areas):
  - **functional** in terms of employment, housing or retail areas or the areas within which people seek jobs or homes (larger area

Box 3.3 Bay of Pasaia (Spain) — integrated urban development across municipal borders

Situation
The Bay of Pasaia lies within the metropolitan area of San Sebastian, which is demographically the second-most important area of the Basque country (Spain) with 240 000 inhabitants in the municipalities of San Sebastián, Rentaría, Lezo and Pasaia. It encompasses a complex area with economic problems, high population densities, environmental damage, as well as port and rail infrastructures intermixed with residential areas. Its municipalities are fragmented, and competences dispersed hampering the application of integrated approaches.

Photo: © URBAN, San Sebastián-Pasaia

Solution
Given this complexity, and building on a broad political consensus through the participation of all the municipalities since 1994, the PIC URBAN II San Sebastián-Pasaia project started in 2001. The project philosophy was ambitious: on the one hand initiating changes which had the greatest chance of success and on the other hand finding new and innovative opportunities for development. To achieve this, the project focused on 4 areas:

- Urban regeneration: new urban planning for the renovation of places and areas, which contribute significantly to the attractiveness of the area, and which add environmental quality to the bay. The focus is on new locations for the development of innovative activities, parks and green areas, pedestrian zones, improved accessibility, and the establishment of cycle paths etc.;
- Creating jobs by developing new economic activities with the support of business: Developing new leisure and culture activities, meanwhile continuing to support new business opportunities based on an adequate financial structure;
- Socio-economic reintegration: countering social exclusion requires a multidisciplinary emphasis that enables the integration of people on the basis of personalised treatment by developing infrastructures expanding social engagement;
- Developing the potential and access to information technologies by improving infrastructures and by personalised training measures.

Results
The improvement of the situation is evident: more than 20 million Euros were invested in new infrastructure, parks, public places etc. There has been an unprecedented increase in the numbers of tourists to the area and the creation of more than 7 000 jobs in the IT area.

Although the full range of achievements can only be assessed from a longer term perspective, the approach adopted has initiated the wide ranging public participation necessary to achieve the renovation and improvement of the region in the near future.

The financial investments necessary to achieve a profound transformation of the region are great. In this respect ERDF URBAN funding has proved invaluable in establishing inter-institutional collaboration involving local governments, the County Council of Gipuzkoa, as well as the Basque and the Spanish governments; a cooperation which focuses on public companies as instruments of implementation.

More information: http://www.bahiadepasaia.com.

than the city or town), or in terms of urban networks, for example transport systems;
- **typology** including the distribution of urban services, different forms of urban society, and variation in population densities;
- **administrative** according to the boundaries of government agencies;
- **morphological** according to the actual area covered by urban land use.

The integrated urban management approach addresses all urban processes, whether they are governed by the city or town administration or other administrative levels including the regional level, state, EU, and global levels. Such an approach includes also city to city, urban-rural and local to global interactions considered from the urban perspective.

- **Integrated**
The development of a holistic perspective on urban management, that considers the various interlinkages within the urban environment, and seeks to combine the related processes in order to develop greater coherence and mutual reinforcement of planned responses to the challenges generated by the key drivers of urban development. Of particular importance is the integration of plan-making and plan-delivery mechanisms.
The different dimensions to be considered include:

- **horizontal integration** between different policy and programme areas including economy, social affairs, environment, culture…;
- **vertical integration** between levels of governance ranging from local to global;
- **spatial connection** of towns, cities, rural hinterland and regions;
- **temporal linkage** of the impacts of current developments in relation to the potentials for future development;
- **balancing** individual and group interests with societal needs.

Integrated management requires appropriate integrated institutional structures and information bases in respect of all the above dimensions.

- **Management**
Getting people, stakeholder groups, business and administrations to act together towards a common goal, for example achieving a certain quality of life. This includes planning, leadership, organisation, resourcing — human, financial, technological as well as natural resources — monitoring and evaluating the process of sustainable urban development to enable corrections and adaptations. It needs to be spatially coherent in order to take the right decisions not only at the right level but in ways that reflect the spatial functionality of Europe.

3.5 Steps towards implementation

All sectors of society and all administrative levels can gain long-term benefits from applying and being part of an integrated approach, creating good governance for urban areas. Instead of cities and towns competing for jobs, tax and other funding, local government can mobilise more resources, creativity and support in delivering desired outcomes and managing unwanted change. However, such an approach requires strong political support. As claimed by the European Parliament (2008), the EU should not only financially support the use of such approaches at national, regional or local level, but also analyse, when providing funding, to what extent a binding requirement is feasible. Equally, it should apply these principles in its own policy-making. This would make funding of local and regional projects more efficient and better enable the integration of supportive measures, such as standard setting and taxation, at EU and national level.

New governance through partnership

As problems can seldom all be solved at one level or within one policy sector, successful implementation of the integrated approach requires the active participation of all actors, which can range from individual citizens to the European Union. Therefore, governmental action needs to shift to new forms of governance through partnership. Integration of individual action programmes with other interdependent areas and administrative levels must become a basic and natural principle of all bodies.

This involves:

- new governance arrangements;
- inter-governmental relationships and connections between areas of concern, regardless of jurisdiction;
- engagement in more coordinated decision-making;
- new partnerships and approaches to action with local organisations and citizens;
- more accountability in fulfilling commitments.

Box 3.4 Effective governance — metropolitan spatial planning for self-assessment (InterMETREX project)

Initial situation
With most of the population of the EU living in urban areas and the need for the harmonious, balanced and sustainable development of metropolitan regions, it is essential to have the capacity to audit the effectiveness of governance of these major regions and the urban areas within them.

Photo: © METREX

Solution
Under the INTERREG IIC and IIC programmes the METREX network developed a set of benchmarks for self-evaluating the effectiveness of strategic planning and governance, based on the principle of self diagnosis and continuous improvement. The adoption of this strategy reflected the fact that no one system suits all urban regions and that no area had an ideal system.

The project identified those factors which could be systematically evaluated and were capable of being detached from political judgements. The first relates to the ability of the planning mechanisms to take effective decisions — i.e. what are their 'powers' or 'competences'? The second relates to the ability of the organisation to take informed decisions — i.e. what are its technical resources or 'capabilities'? The third relates to the ability to make decisions which are accepted by those affected by them — i.e. what opportunities are provided for engagement in the planning processes'?

Results
The project concluded that an incremental approach was probably required but that there are common issues which help explain the differences in the effectiveness of city regions including the following:

- The coherence of the areas as a Functional Urban Region (FUR); More coherent areas in terms of social economic geography are less dependent upon decisions taken by adjoining areas and have greater scope for resolving conflicts locally.
- The ability to deliver large projects, both in terms of financial resources and organisational skills: Large projects are often introduced for the transformation or re-engineering of a region's infrastructure and for stimulating public interest in the plan. If, however, they are not deliverable this can generate blight and loss of credibility in governance generally.
- The existence or not of a national planning framework: Increasingly, local decisions are dependent upon national policy and commitment.
- Planning for uncertainty: Strategic Planning seeks to set out the vision to provide longer term confidence, to safeguard the interest of communities affected by change, and for those who risk their investment in new development. The practical response to this dilemma is a commitment to phasing, monitoring and to a five-year review.
- The available technical capacities: the lack of effective technical and professional capacities undermines the credibility and deliverability of EU goals for sustainable cities.

As a concrete result, the experience gained over the course of the project was summarized and transformed to deliver the guide 'METREX Practice Benchmark of Effective Metropolitan Spatial Planning' which contains 25 benchmarks of effective competence, capability and process. In the form of a checklist, the guide enables metropolitan regions and areas to easily self assess their current practice, identify potentials for improvements, and plan further steps to achieve better governance.

More information: http://www.eurometrex.org/ENT1/EN/Activities/activities.asp.

The new forms of governance need to improve the linkage of stakeholders to policy processes, through consensus-building, participation and coordination. Instruments for supporting governments in evaluating their sustainability processes are already well-established ([14]), with practical expert-led management complemented by bottom-up community visioning (Box 3.4).

Involvement and participation of the different levels and stakeholders should be continuous and vary according to the requirements of the integrated management cycle. Participatory decision-making is desired and demanded by citizens who wish to play a more active role in the governance of their society. Regional, national and local governments, NGOs, as well as the scientific community and business interests are also increasingly eager to reap the benefits of engaging actively in decision-making processes. Enabling wide participation ensures the acceptance and sustainability of policy implementation.

Good governance in relation to vertical integration requires the reinterpretation of the subsidiarity principle. Section 2.1–2.6 of this report have demonstrated that major urban problems cannot be solved at only one administrative level. Responsibilities need to be defined in relation to the many interlinkages between European, national and local policy (for example Box 3.5). The EU needs an urban approach that is neither a new policy area nor a top down 'one size fits all' administrative process, but an auditing of the impacts of EU policy in terms of their implications for the urban level. The EU also needs to develop supportive cross-sectoral policies for urban areas.

Long-term vision

Integrated approaches and good governance need long-term strategic visions — for example as recognised in the Guidance to the *Thematic Strategy on the urban environment* (EC, 2007g). The different actors involved in urban development need a shared vision of the quality of life to be maintained or attained. A common vision is an indispensable prerequisite that links the different policies at different administrative levels, and facilitates the delivery of coherent actions. Common vision can also reduce pressures on sectoral policies to act in a short-term timeframe in order to produce immediate success that is almost certainly not sustainable. With a shared long-term vision, policy can demonstrate its ability to fulfil the vision, and can also justify actions that will be successful in the long term, and only in combination with other partner actions. The potential for local councils to lead the debate and promote change is well illustrated by the London Climate Change Action Plan (see Box 3.6).

Management cycle

A crucial aspect of successful integrated urban management is the application of a cyclical approach, consisting of five major steps that are repeated in regular cycles, according to the specific circumstances. A baseline review documenting the current environmental and administrative situation, legal requirements and political priorities prepares the ground. On this basis, objectives and targets are discussed, agreed, set and approved and actions and initiatives are identified according to current technologies and life styles. The timeframes related to these targets provide for future monitoring, review and evaluation of the process.

New information permits the validation of established policies, and if necessary for new decisions are taken, and the cycle recommences. Once the mechanism is established, in subsequent years the entire process is repeated. All the above steps are linked in a continuous process; the targets set are (re)defined as an essential element of government procedure with increasing effectiveness and coherence. The European project MUE 25 further defines such an approach (Box 3.7).

Improved data and knowledge

Policy-makers need a solid basis of information and intelligence to support decision-making. There is a need to organise information on urban development in a consistent and integrated way to support integrated policy-making, not just at the local level, but also interlinked to urban-relevant data at higher administrative levels. For example, at the European level Eurostat collects socio-economic and some environmental data for around 300 cities via the Urban Audit database. The Corine Land Cover project of the European Environment Agency produces land-use maps from satellite images and the ESPON programme provides data on urban functional zones. Integrating such information, complementing it, and linking it to other regional and local data are vital to support assessments and projections of the impacts of urban development, in order to support integrated urban policy-making.

([14]) www.localsustainability.eu; www.localevaluation21.org.

Box 3.5 Participation in European transport policy — Magistrale für Europa

Situation

European and international accessibility is seen as an important precondition for economic growth, European cohesion and international competitiveness of cities and regions. Thus the aim of the European transport policy is to improve accessibility by building the Trans-European Transport networks (TEN-T). However, even if the cities along the routes are substantially affected by the projects, they are not formal partners in their planning and implementation. This is a particular problem as the TENs will only generate their benefits to cities and regions if the European network is complemented by appropriate local and regional infrastructure, transport and spatial planning. The example of the Magistrale für Europa tells a different and encouraging story:

Map 3.1 Main line for Europe

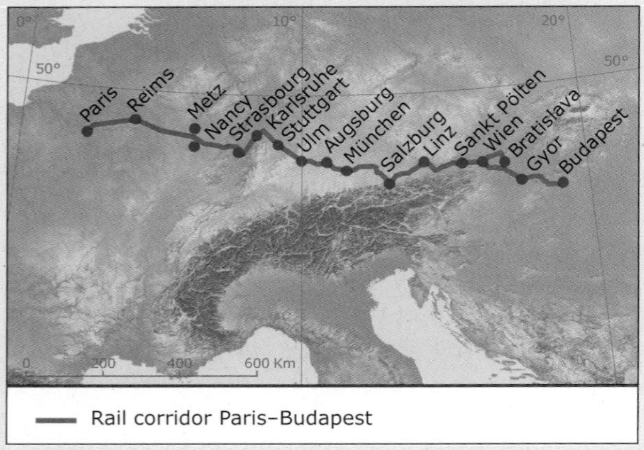

Rail corridor Paris–Budapest

Source: EEA, 2009.

Solution

Around 30 cities and some regional organisations along the Paris-Straßburg-Karlsruhe-Stuttgart-München-Salzburg-Wien-Budapest rail corridor had noticed the advantages of a better international accessibility and founded the alliance Magistrale für Europa with Karlsruhe as the managing city as early as 1990. The aim was to attract the attention of European and national authorities to local interests, to participate in the process and to push for its implementation. This transnational, intercity cooperation is based on regular information exchange, joint opinion formation and lobbying national and European authorities. These activities are accompanied by technical studies, workshops as well as public communication.

Major local engagement led to European acceptance of the requests expressed in the TEN project No 17 in 2004. However, given the inadequate level of coordination among the important actors, implementation of such a cross-border project was difficult and was delayed. As a result, in 2005, the EU assigned a European coordinator for the project, who coordinated the different national authorities and rail companies and also involved the Magistrale für Europa. All sides could now benefit: the EU gained from the comprehensive regional and local knowledge and the local engagement to integrate the TEN project into regional and local infrastructure, and the cities gained much greater benefits by influencing the TEN project. As a result, implementation advanced.

Results

As a consequence of its long-term activities, the city alliance found a way to influence European transport policy to their benefit and to participate in the TEN project, which was not formally foreseen in the process. The fact that the alliance is still active after 18 years and that the partners finance a joint secretariat demonstrate that the cities value their direct engagement in European transport policy as very effective and beneficial for them.

Concrete local or regional actions, which are not part of the TEN project but for which integration is absolutely necessary are, for instance, the city stations which serve as intermodal intersections for European and local transport and compatible timetables. For example, Vienna has different train terminuses serving east and west so passengers travelling between Paris and Budapest need to take public transport to travel from one to the other to proceed with their journey. The construction of a new central station and a tunnel has been agreed among the Austrian government, the City of Vienna and the Austrian Railways (ÖBB) enabling continuous trips, major time savings and greater convenience in the future.

More information: www.magistrale.org.

Source: Ismaier and Seiß, 2005.

Box 3.6 London Climate Change Action Plan (the United Kingdom) — clear vision and concrete targets

Photo: © EEA

How local integrated action can make a European and global difference: the London Climate Change Action Plan

Initial situation

London produces 8 % of the CO_2 emissions of the United Kingdom which in turn is the world's eighth largest emitter of these emissions. Without any action London's emissions will increase even further in the future. Stabilising the global carbon emissions at a level where catastrophic climate changes are avoided will require enormous emission reductions throughout the developed world. As the challenge is huge, a clear vision, concrete targets, and strong political leadership are absolutely necessary.

Solution

Prior to the creation of the Greater London Authority (GLA) there was no strategic capacity to plan and manage change in one of the world's major cities. The GLA brought together economic, developmental, transport and environmental planning for eight million people under the leadership of a single mayor accountable to an elected assembly. Following several actions from the year 2000 onwards, the GLA produced a 'Climate Change Action Plan' in February 2007 with the specific aim 'to deliver decisive action in London with the urgency that is required' to tackle the potential threats to London of Climate Change. 'The Mayor's new target for London, therefore, is to stabilise CO_2 emissions in 2025 at 60 % below 1990 levels, with steady progress towards this over the next 20 years. As part of London's ambitious target 30 % is to be achieved within the responsibility of the GLA and the other 30 % as part of national government action such as decarbonising the electricity grid. This target is considerably more ambitious than the UK government's current aspiration of a 60 % reduction from 2000 levels by 2050' (the new UK government target is now 80 % by 2050).

The plan focuses on the next 10 years in the context of achieving the 2025 target. It comprises all CO_2 relevant urban activities and lists many concrete measures and targets in its different actions and programmes, such as The Green Homes Programme, Green Organisations Programme, Energy Efficiency Programme and Requirements for new developments, Transport Related Programmes, and the Delivery Mechanisms.

Results, lessons learnt and transfer potential

The adoption of such ambitious targets and development of concrete measures is a success in itself, in particular, as the newly elected city government has endorsed this plan and is on the way to propose revised implementation measures. In part this achievement arises as a result of raising general awareness over the last years due to intensive work by the administration and others in the United Kingdom and, in particular, the broad partnership-approach including the private sector. The clear target and the Action Plan help to combine and streamline the efforts of each partner, thereby reaching the necessary effectiveness. These partnerships will be crucial for London to deliver and achieve its targets especially in the light of governmental changes, energy and financial crisis.

With its ambitious approach, London has inspired others and has taken a political lead on climate change among large cities; for example, in the C40 Large Cities climate leadership group. Consequently, the results of London's Climate Change policy stretch across Europe and the world.

More information: http://www.london.gov.uk/mayor/environment/climate-change/ccap/index.jsp.

Box 3.7 ManagingUrbanEurope 25 project (MUE 25) — integrated urban management

The ManagingUrbanEurope-25 project (MUE 25) developed guidelines for cities and regions on how to coordinate the efforts of public and private sectors as well as a cyclical procedure for an integrated management.

The system should be developed in accordance with the following basic principles:

Relevance: addressing the needs of all relevant activities and actors. It must also address key issues facing cities, common problems and common solutions, with potential for engagement with key issues facing all European cities.

Functional perspective: the system should address the urban area, irrespective of administrative boundaries and degree of local authority power, including the impact of activities of all actors (municipality and stakeholders) on neighbouring municipalities and cities.

Legal compliance: the system needs to assist urban area's legal compliance.

Figure 3.1 The sustainability cycle

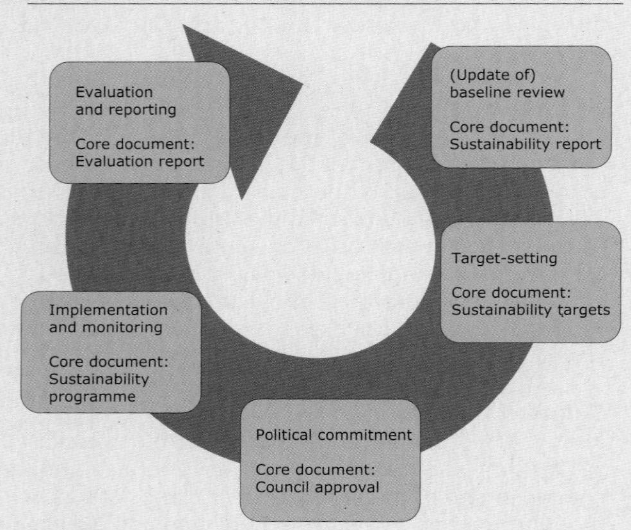

Source: ICLEI — Local Governments for Sustainability.

Continuous improvement towards sustainability: the system needs to assist the urban area's continual measurable improvement towards sustainability. To this end, it has to have a periodic and cyclical nature.

Strategic orientation: the system has to be considered as a mechanism to inform decision-making and support implementation. To this end, it has to focus on strategic rather than operational issues.

Mainstreaming: the system has to be organised centrally in the city management. Regular involvement of the central political body in target setting and evaluation will ensure political commitment, legitimisation and maximised impacts. The process is subject to continuous review and assessment on an annual basis in line with the prime annual budget cycles.

Decentralised implementation: the coordination of the system has to be based within the local administration. However, the strategic goals and targets are to be operationalized and implemented via a range of actors including administrative departments, private companies and relevant stakeholders. The system needs to allow for the derivation of specific goals and targets for these using existing (sectoral) instruments, such as land-use planning, air quality management, water quality management, transport planning etc.

Integration: the system will ensure horizontal integration across various departments and engagement with all relevant stakeholders in the city, and vertical integration by addressing local regional and national spheres of government.

Inclusive: the system will allow for appropriate involvement of urban stakeholders and provide for transparency and communication in decision-making and evaluation.

Adaptability: the management system has to be adaptable to variations in local contexts, as cities are different in size, economic level, organisation, and the activities they pursue.

Complementary: the urban integrated management system will not replace existing and applied environmental management instruments in cities, but build on them, as well as coordinate and integrate existing (sectoral) instruments.

Box 3.7 ManagingUrbanEurope 25 project (MUE 25) — integrated urban management (cont.)

Evolutionary: the system will build on existing experience with environmental management systems rather than re-invent the wheel.

Gradual expansion: the cities can gradually expand the system in scale and scope to include various aspects, actors and spheres of government. Through the integration of social and economic dimension the urban integrated management system will develop to include all sustainability dimensions in the management system.

Source: http://intra.mue25.net/, http://www.localsustainability.eu/.

4 Summary and conclusions

Quality of life in cities and towns is more than a local concern. Urban areas in Europe accommodate nearly 75 % of the population and generate a substantial ecological footprint that not only impacts on their own ability to generate a high quality of life, but also on that of their immediate rural hinterlands in Europe and globally.

Quality of life at risk

Cities are the places where quality of life is experienced and also generated. Over the past decades, urban quality of life has substantially improved; yet, in society at large, ground is being lost: serious health problems are developing arising from air pollution and noise, the number of obese people is increasing, and major economic, environmental and social impacts are foreseen as a consequence of climate change. The problems are serious, and we are on the brink of potentially irreversible change. While our current way of life provides us with quality of life, at the same it is putting our future at risk. A change towards more sustainable life styles, but which nonetheless provide all necessary satisfaction and happiness, is required, and policy must set the frame.

Barriers to policy implementation

In general, the management instruments necessary to cope with the challenges ahead are available, the first practical experience has been secured, and further innovative solutions are being developed. However, a broad implementation plan is still is lacking. The main barriers to implementation are listed below:

- sectoral and short-term policy-making in an attempt to secure rapid results is still the norm; however, it is clear that the size and complexity of current problems require cross-sectoral, cooperative and long-term approaches, which do not typically deliver results in the short-term;
- technical improvements of current processes alone will help but not solve the problems faced. It is also necessary to modify urban life styles, the way we fulfil our needs and demands, and embrace new alternatives;
- although cities play a crucial role in securing quality of life, local–European partnerships still need to be developed. A strict interpretation of the subsidiarity principle limits the search for solutions to single administrative levels, yet today's problems cannot be solved at one level alone.

Integrated partnership approaches

Shared responsibilities and long-term perspectives are necessary to ensure quality of life for all. Such approaches must be consistent across policy levels and sectors and be spatially and socially coherent. All can benefit from such integrated partnerships: cities by getting support from EU urban policy; Europe in securing local actions that are complementary to European action; and surrounding rural areas and regions in ensuring the full representation in strategic decision-making for the locality. Political will allied with a new understanding of the role of cities in the management of complex systems and supported by improved forms of governance will permit the realisation of the full benefits of the integrated urban management approach.

Today, as the instruments are available, it is up to the responsible actors and agencies at all governmental levels to take action and work together with business and citizens on deploying management tools in order to fill the gaps and missing links in knowledge and management supported by creative and cooperative approaches.

Cities and towns

Urban areas need to provide for their citizens the foundations for choices leading towards more sustainable life styles, such as affordable housing in more compact urban areas that provide high quality public spaces and a healthy environment. To ensure that regional, national and European governments are fully supportive of these transformations in lifestyles, cities also need to become active and cooperative partners in applying integrated approaches in collaboration with other levels of governance as well as with other cities and rural areas.

The European Union

The EU needs to set the framework conditions supporting national, regional and local governments. European policy needs consistency in its urban approach based on an audit of the impacts of EU policy in terms of the implications for the urban level and by developing supportive cross-sectoral policies for the urban area. To achieve this new governance, closer partnership with the local level is key, whilst respecting the spatial functionality of Europe and respective responsibilities.

National and regional governments

At the national and regional level governments need to further develop the framework conditions, including legislation, to ensure that urban policy can fully deliver the intended results. National and regional governments can determine, for example, the right environmental price for goods and services; and these governments, in particular, have the power to enable cities and towns to participate in national and thereby European policy-making, thus to ensure more consistent policy across all levels.

Partnership succeeds

The potential of cities and towns to successfully respond to current and future challenges is immense, provided a partnership of all administrative levels is created. A major success of this report has been the extent to which the collaborating urban network partners have been able to integrate their different perspectives and visions of the future of urban Europe. The dialogue deepened the understanding of all partners, built trust, and created a platform for further, better cooperation. The process of developing the report was often not easy, but the results are a rich, multi-perspective analysis and new ideas on how to proceed. This process is itself an example of the success of the desired broad partnership approach.

References

Agència d'Energía de Barcelona, 2006. *El comptador: Informació energètica de Barcelona — núm. 2-2006*, Barcelona.

Ambiente Italia, 2006. *Urban Ecosystem Europe*. Milan.

Ambiente Italia, 2007. *Urban ecosystem Europe: an integrated assessment on the sustainability of 32 European cities*. Milan.

Andrews, FM. and Withey, S. B., 1976. *Social indicators of well-being: American's perceptions of life quality*, Plenum Press, New York.

APAT, 2007. *IV APAT Annual report on urban environmental quality 2007 edition*. Rome.

Arends, B. G., 2005. *Results of a URBIS model run for Rotterdam, calculated by DCMR*.

Babisch, W., 2006. *Transportation noise and cardiovascular risk: review and synthesis of epidemiological studies: dose-effect curve and risk estimation*. Dessau, Umweltbundesamt, WaBoLu-Hefte.

Barredo J. I.; Lavalle, C. and de Roo, A., 2005. *European flood risk mapping*. EC DG JRC, 2005 S.P.I.05.151.EN.

BBR — Bundesamt für Bauwesen und Raumordnung (publisher), 2005. *Gewerbeflächenmonitoring: Ein Ansatz zur Steigerung der Wettbewerbsfähigkeit des regionalen Gewerbeflächenpotenzials in Ostdeutschland*. Bonn.

Berlin-Institut, 2008. *Europe's demographic future: growing regional imbalances*. Earthprint Library.

Boeft, J. Den, 2003. *Aansluiting A4 op Kethelplein en het plan IODS: aspect luchtkwaliteit*. TNO R 2003/326.

BMVBS Bundesministerium für Verkehr, Bau und Stadtentwicklung, 2006. *kommKOOP: Erfolgreiche Beispiele interkommunaler Kooperationen*. Dokumentation des MORO-Wettbewerbs 2005/2006. Bonn.

Bretagnolle, A.; Paulus, F. and Pumain, D., 2002. 'Time and space scales for measuring urban growth' [online], *Cybergeo, July 25 2002*. www.cybergeo.eu/index3790.html [accessed April 2009].

BPL, 2005. *Effecten van klimaatverandering in Nederland*. MNP-rapportnummer: 773001034. Bilthoven: MNP.

CAFÉ, 2005. *Thematic strategy on air pollution*. COM(2005) 446 final.

Campbell, A.; Converse, P. E. and Rodgers, W. L., 1976. *The quality of American life: perceptions, evaluations, and satisfactions*. Russell Sage Foundation.

Champion, A. and Hugo, G., 2004. *New forms of urbanization: beyond the urban-rural dichotomy*. Ashgate Publishing Ltd., Aldershot, United Kingdom.

CBS-Statistics Netherlands, 2004. *Statistisch jaarboek 2004*. Den Haag, the Netherlands.

Cuadrat Prats, J. M.; Vicente-Serrano, S. M. and Saz Sánchez, M. A., 2005. 'Effectos de la urbanización en el clima de Zaragoz (Espana): la isla de calor y sus factores condicionantes'. *Boletin de la A.G.E. N.* 40: pp. 311–327.

Dankers, R. and Hiederer, R., 2008. *Extreme temperatures and precipitation in Europe: analysis of a high-resolution climate change scenario*. EUR 23291 EN. Office for Official Publications of the European Communities Luxembourg.

Dicken, P., 2004. Global shift: industrial change in a turbulent world. *Progress in Human Geography* 28: pp. 507–515.

EC — European Commission, 2006a. *The demographic future of Europe: from challenge to opportunity*. Luxembourg.

EC — European Commission, 2006b. *Community strategic guidelines on cohesion*. Council Decision of 6 October 2006 (2006/702/EC). OJEU L 291, 21.10.2006.

EC — European Commission, 2006c. *Cohesion policy and cities: the urban contribution to growth and jobs in the regions*. COM(2006) 385 final.

EC — European Commission, 2006d. *Thematic Strategy on the urban environment.* COM(2005) 718 final.

EC — European Commission, DG Regional Policy, 2007a. *Survey on perceptions of quality of life in 75 European cities*. Brussels.

EC — European Commission, DG Regional Policy, 2007b. *State of European cities report.* Brussels.

EC European Commission, 2007c. *Europe's demographic future: facts and figures on challenges and opportunities.* Luxembourg.

EC — European Commission, 2007d. *Green Paper: Towards a new culture for urban mobility*. COM(2007) 551 final. Brussels 25.9.2007.

EC — European Commission, 2007e. EU research on environment and health: results from projects funded by the Fifth Framework Report, Luxembourg: Office for Official Publications of the European Communities.
EC — European Commission, 2007f. Green Paper: Adapting to climate change in Europe: options for EU action. COM(2007) 354 final. Brussels 29.6.2007.

EC — European Commission, 2007g. *Integrated environmental management: guidance in relation to the Thematic Strategy on the Urban Environment*. Luxembourg: Office for Official Publications of the European Communities. www.ec.europa.eu/environment/urban/home_en.htm [accessed April 2009].

EC — European Commission, 2007h. *The urban dimension in Community policies for the period 2007–2013*. Brussels. www.ec.europa.eu/regional_policy/sources/docgener/guides/urban/index_en.htm [accessed April 2009].

EC — European Commission, 2007i. *Fourth report on economic and social cohesion*. COM (2007) 273 final. Brussels, 30.5.2007.

EC — European Commission, 2008a. *Climate change and international security*. Paper from the high representatives and the European Commission to the European Council (7249/08). Brussels.

EC — European Commission, 2008b. *Fostering the urban dimension: analysis of the operational programmes co-financed by the European Regional Development Fund* (2007–2013). Working document of the Directorate-General for Regional Policy. Brussels 25 November 2008.

EC — European Commission, 2008c. *Green Paper on territorial cohesion: turning territorial diversity into strength*. COM(2008) 616 final. Brussels, 6.10.2008.

EC — European Commission, 2009. *White Paper: Adapting to climate change: Towards a European framework for action*. COM(2009) 147 final. Brussels, 1.4.2009.

EEA — European Environment Agency, 2005. *Effectiveness of urban wastewater treatment policies in selected countries: an EEA pilot study*. EEA Report No 2/2005. Copenhagen.

EEA — European Environment Agency 2006a. *Urban sprawl in Europe: the ignored challenge*. EEA Report No 10/2006. Copenhagen.

EEA — European Environment Agency, 2006b. *Transport and environment: facing a dilemma — TERM 2005.* EEA Report No 3/2006. Copenhagen.

EEA — European Environment Agency, 2007a. *CSI 004: Exeedances of air quality limit values in urban areas* (version 2). http://themes.eea.europa.eu/IMS/CSI. Copenhagen.

EEA — European Environment Agency, 2007b. *Europe's environment — the fourth assessment*. Copenhagen.

EEA — European Environment Agency 2008. *Impacts of Europe's changing climate: 2008 indicator-based assessment*. EEA Report 4/2008. Copenhagen.

EEA — European Environment Agency, 2009a (forthcoming). *Report on environmental impacts of European consumption and production patterns*. Copenhagen.

EEA — European Environment Agency, 2009b (forthcoming). *Territorial cohesion: analysis of environmental aspects of EU cohesion policy in selected countries*. Copenhagen.

Ellaway, A.; Macintyre, S. and Xavier, B., 2005. 'Graffiti, greenery and obesity in adults: secondary analysis of European cross sectional survey'. *British Medical Journal* 331: pp. 611–612.

ENEA — European Network of Environmental Authorities, 2006. *The contribution of Structural and Cohesion Funds to a better environment*. Luxembourg.

EPI — Earth Policy Institute, 2006. *Setting the record straight — more than 52,000 Europeans died from heat in summer 2003*. Washington. www.earth-policy.org/Updates/2006/Update56.htm [accessed April 2009].

ESPON — European Spatial Planning Observational Network, 2004. *ESPON 1.2.1: Transport services and networks: territorial trends and basic supply of infrastructure for territorial cohesion*. University of Tours — lead partner. www.espon.eu/mmp/online/website/content/projects/259/652/index_EN.html [accessed April 2009].

ESPON — European Spatial Planning Observational Network, 2005a. *ESPON 1.1.1: Potentials for polycentric development in Europe*. www.espon.eu/mmp/online/website/content/projects/259/648/index_EN.html [accessed April 2009].

ESPON — European Spatial Planning Observational Network, 2005b. *EPSON 1.1.4: The spatial effects of demographic trends and migration*. www.espon.eu/mmp/online/website/content/projects/259/651/index_EN.html [accessed April 2009].

Eurobarometer, 2005. *Urban audit perception survey: local perception of quality of life in 31 European cities*. Flash Eurobarometer 7/2005. www.ec.europa.eu/public_opinion/flash/fl_156_en.pdf [accessed April 2009].

Eurobarometer, 2008a. *Attitudes of European citizens towards the environment* (in French). http://ec.europa.eu/public_opinion/archives/ebs/ebs_295_fr.pdf [accessed April 2009].

Eurobarometer, 2008b. *Europeans' attitudes towards climate change*. http://ec.europa.eu/public_opinion/archives/ebs/ebs_300_full_en.pdf [accessed April 2009].

Eurofound — European Foundation for the Improvement of the Living and Working Conditions, 2004. *Quality of life in Europe, first European quality of life survey 2003*. Dublin. www.eurofound.europa.eu/pubdocs/2004/105/en/1/ef04105en.pdf [accessed April 2009].

Eurofound — European Foundation for the Improvement of the Living and Working Conditions, 2007. *First European quality of life survey: key findings from a policy perspective*. Dublin.

Eurofound — European Foundation for the Improvement of the Living and Working Conditions, 2008. *Second European quality of life survey — first findings*. Dublin.

European Parliament, Committee on Regional Development, Kallenbach, G., 2008. *Report on the follow-up of the territorial agenda and the Leipzig Charter: towards a European action programme for spatial development and territorial cohesion*.

Eurostat/IFF, 2007. *Economy-wide Material Flow Accounts, Resource Productivity, EU-15 1990–2004, Environmental Accounts*. Helga Weisz, Willi Haas, Nina Eisenmenger, Fridolin, Krausmann, Anke Schaffartzik. Vienna.

Eurostat, 2007. *EU economic data pocketbook, 2-2007, Quarterly*. ISSN 1026-0846, European Commission. Luxembourg.

Eurostat, 2008a. *Population projections 2008–2060 — from 2015, deaths projected to outnumber births in the EU 27*. Luxembourg, 26 August 2008.

Eurostat 2008b. *Tourism in Europe: does age matter?* Statistics in focus 69/2008. http://epp.eurostat.ec.europa.eu/cache/ITY_OFFPUB/KS-SF-08-069/EN/KS-SF-08-069-EN.PDF [accessed April 2009].

FAOSTAT — Statistical database of the UN Food and Agriculture Organisation, 2003. http://faostat.fao.org. [accessed April 2009].

Gjestland, T., 2007. 'The socio-economic impact of noise: A method for assessing noise annoyance'. *Noise Health* 9: pp. 42–44.

Hacker, J. N.; Belcher, S. E. and Connell, R. K., (UKCIP), 2005. *Beating the heat: keeping UK buildings cool in warm climate*. UKCIP Briefing Report. UKCIP, Oxford www.ukcip.org.uk.

Haskoning, R., 2008. *Spatial planning in the context of nature-oriented flood damage prevention: comparison for the NWE countries*.

Health Survey for England, 2003. National Centre for Social Research, Department of Epidemiology and Public Health at the Royal Free and University College Medical School. Commissioned by Department of Health.

Hideg, G. and Manchin R., 2007. *Environment and safety in European capitals*. EU ICS working papers. February 2007.

Hiederer, R. and Lavalle, C., 2009. *Geographic Position of Europe for End-of-Century Temperature Equivalent*. Special Publication JRC Pubsy N. 50603, European Communities.

Hiederer, R.; Dankers, R. and Lavalle, C., 2009. *Evaluating Floods and Heat Wave Hazards for a Climate Change Scenario*. Special Publication JRC Pubsy N. 50566, European Communities.

High Level Group, 2003. *On the trans-European transport network* [online]. www.ec.europa.eu/ten/transport/revision/hlg/2003_report_kvm_en.pdf [accessed April 2009].

Hoek, G.; Brunekreef, B.; Goldbohm, S.; Fischer, P. and van den Brandt, P. A., 2002. 'Association between mortality and indicators of traffic-related air pollution in the Netherlands: a cohort study'. *The Lancet* 360: pp. 9341–47.

Hübner, D., 2006. 'Cities for growth, jobs and cohesion: the urban action of the Structural Funds'. speech in *Inforegio num.* 19 April 2006.

Hübner, D., 2008a. Knowing and designing our cities to cope with today's and tomorrow's challenges. Speech at the Urban Audit conference, Brussels, 20 June 2008.

Hübner, D., 2008b. 'Control of structural actions — meeting the challenge'. Speech at the seminar Control of structural actions, Brussels, 10 June 2008.

ICLEI — Local Governments for Sustainability, 2002. *Second local agenda 21 survey.* http://www.iclei.org/index.php?id=1185 [accessed April 2009].

IEA — International Energy Agency, 2008. *World Energy Outlook 2008.* Geneva.

IPCC, 2007. *Climate change 2007: synthesis report contribution of Working Groups I, II and III to the Fourth Assessment Report of the Intergovernmental Panel on climate change.* Cambridge University Press, Cambridge, United Kingdom.

Ismaier, F. and Seiß, R., 2005. Europäische Verkehrspolitik von unten — Die 'Magistrale für Europa': ein transnationales Bündnis für die Schiene. *Raumplanung* 118 (2005). pp.177–181.

Jarup, L.; Babisch, W.; Houthuijs, D.; Pershagen, G.; Katsouyanni, K.; Cadum, E.; Dudley, M-L.; Savigny, P.; Seiffert, I.; Swart, W.; Breugelmans, O.; Bluhm, G.; Selander, J.; Haralabidis, A.; Dimakopoulou, K.; Sourtzi, P.; Velonakis, M. and Vigna-Taglianti, F. on behalf of the HYENA study team, 2008. 'Hypertension and exposure to noise near airports: the HYENA study'. *Environmental Health Perspectives* Volume 116, Number 3, March 2008: pp. 329–333.

Kavanagh, A. M.; Goller, J. L.; King, T.; Jolley, D.; Crawford, D. and Turrell, G., 2005. 'Urban area disadvantage and physical activity: a multilevel study in Melbourne, Australia'. *Journal of Epidemiology and Community Health* 59 (11): pp. 934–940.

van Kempen, E. E. M. M, 2008. *Transportation noise exposure and children's health and cognition.* Doctoral thesis Utrecht University. http://igitur-archive.library.uu.nl/dissertations/2008-0122-203944/index.htm [accessed April 2009].

Kern, K. and Bulkeley, H., 2009. 'Cities, Europeanization and multi-level governance: governing climate change through transnational municipal networks, transnational municipal networks'. *Journal of Common Market Studies* 47(2).

Lavalle, C.; Barredo, J.; Mc Cormick N.; Engelen, G.; White, R. and Uljee, I., 2004. *The MOLAND Model for Urban and Regional Growth Forecast. A tool for the Definition of Sustainable Development Paths.* EUR 21480 EN. Office for Official Publications of the European Communities Luxembourg.

Lipovetsky, G., 2006. *Le bonheur paradoxal.* Essai sur la société d'hyperconsommation, Paris, Gallimard.

Lois-González, R. C., 2004. 'A model of Spanish–Portuguese urban growth: the Atlantic axis'. *Dela* 21: pp. 281–294.

Miljödepartementet, 2006. *Översvämningshot. Risker och åtgärder för Mälaren, Hjälmaren och Vänern* [Threat of Flooding. Risks and Measures for Mälaren, Hjälmaren and Vänern]. Delbetänkande av Klimat- och sårbarhetsutredningen, SOU 2006:94. Stockholm: Statens offentliga utredningar.

Massam, B. H., 2002. 'Quality of life: public planning and private living'. *Progress in Planning* 58 (3):pp. 141–227.

OECD — Organisation for Economic Co-operation and Development, 2001. *Ageing and transport.*

PBL — Netherlands Environmental Assessment Agency, 2008. *Urbanisation dynamics and quality of place in Europe,* EURBANIS report 1. Planbureau voor de Leefomgeving (NEAA), Bilthoven, the Netherlands.

PBL — Netherlands Environmental Assessment Agency, 2009. *Quality of place in selected European cities,* EURBANIS report 2. Planbureau voor de Leefomgeving (NEAA), Bilthoven, the Netherlands.

Porter, M. E., 1990. *The competitive advantage of nations,* London: Macmillan.

POST — Parliamentary Office of Science and Technology, 2000. *Water efficiency in the home.* Note 135, London.

van Ravesteyn, N. and Evers, D., 2004. *Unseen Europe: a survey of EU politics and its impacts on spatial development in the Netherlands.* The Hague: Netherlands Institute for Spatial Planning.

Rifkin, J., 2000. *The age of access,* Tarcher/Putnam.

RIVM — Dutch National Institute for Public Health and the Environment, 2004. *Hinder door milieufactoren en de beoordeling van de leefomgeving in Nederland Inventarisatie verstoringen 2003.* Franssen, E. A. M.; van Dongen, J. E. F.; Ruysbroek, J. M. H.; Vos, H. and Stellato, R. K. (Eds.). TNO rapport 2004–34. RIVM rapport 815120001/2004.

Rosamund, B., 2004. 'The new theories of European integration' in Cini, M. (ed.): *European Union Politics.* Oxford University Press.

Sagris, V.; Kasanko, M.; Genovese, E. and Lavalle, C., 2006. *Development Scenarios for Eastern European Cities and Regions in the New Europe.* In Conference Proceedings: Proceedings of the 46th Congress of the European Regional Science Association. Volos (Greece): University of Thessaly, Department of Planning and Regional Development. JRC33441.

Sen, 2003. 'Development as Capability Expansion' in *Readings in Human Development*, S. Fukuda-Parr *et al.*, eds. New Delhi and New York: Oxford University Press.

Stansfeld, S. A.; Berglund, B.; Clark, C.; Lopez-Barrio, I.; Fischer, P.; Öhrström, E.; Haines, M. M.; Head, J.; Hygge, S.; van Kamp, I. and Berry, B. F. on behalf of the RANCH study team, 2005. 'Aircraft and road traffic noise and children's cognition and health: a cross-national study'. *The Lancet*, Volume 365, Issue 9475.

Stead, D. and Marshall, S., 2001. 'The relationships between urban form and travel patterns: an international review and evaluation'. *European Journal of Transport*, 1, no. 2, pp. 113–141. www.ejtir.tudelft.nl/issues/2001_02/pdf/2001_02_01.pdf [accessed April 2009].

Stern, N., 2006. *The Stern review: the economics of climate change*. Cambridge University Press http://www.hm-treasury.gov.uk/sternreview_index.htm [accessed April 2009].

Sustainable Development Commission UK, 2008. *Health, place and nature: how outdoor environments influence health and well-being: a knowledge base* http://www.sd-commission.org.uk/publications/downloads/Outdoor_environments_and_health.pdf [accessed April 2009].

UBA — Umweltbundesamt, 2005. 'Data on the environment'. *The state of the environment in Germany*. Dessau: Federal Environmental Agency, pp. 78–92.

UBA — Umweltbundesamt, 2007. *Sozialdemographischer Wandel in Städten und Regionen — Entwicklungsstrategien aus Umweltsicht*. UBA-Texte 18/2007. Dessau.

UBA — Umweltbundesamt (publisher), 2009. *Das Kostenparadoxon der Baulandentwicklung — Von der Außen- zur Innenentwicklung in Städten und Gemeinden* (publication in progress). Dessau-Roßlau.

UITP — International Association of Public Transport, 2006: *Mobility in cities database*.

UK Office for National Statistics, 2007. *Family spending — a report on the family expenditure survey.* http://www.statistics.gov.uk/.

UN — United Nations, 2008. *World urbanization prospects — the 2007 revision*. New York.

UNECE, 2006. *Bulletin of housing statistics for Europe and North America 2006.* www.unece.org/hlm/prgm/hsstat/Bulletin_06.htm [accessed April 2009].

UNEP — United Nations Economic Commission for Europe, UN-HABITAT, 2008. Launch publication *Local action for biodiversity*.

Velarde, M. D.; Fry, G. and Tveit, M., 2007. 'Health effects of viewing landscapes: landscape types in environmental psychology'. *Urban Forestry & Urban Greening* 6: pp. 199–212.

WECD — World Commission on Environment and Development, 1987. *Our common future, report of the World Commission on Environment and Development.* Published as Annex to General Assembly document A/42/427.

Wendel-Vos, G.; Schuit, A.; De Niet, R.; Boshuizen, H.; Saris, W. and Kromhout, D., 2004. 'Factors of the physical environment associated with walking and bicycling'. *Medicine and Science in Sports and Exercise* 36: pp 725–730.

WWF — World Wildlife Fund for Nature, 2008. *Living Planet Report 2008*. Gland 2008.

WWF-UK — World Wildlife Fund for Nature-UK, Carbon Plan, 2007. *Ecological footprint of British city residents*. Godalming.